T0400319

MicroRNA Interference Technologies

Zhiguo Wang

MicroRNA Interference Technologies

Dr. Zhiguo Wang
Montreal Heart Institute
Research Center
5000 Belanger Street
Montreal QC H1T 1C8
Canada
zhiguo.wang@icm-mhi.org
or
wz.email@gmail.com

ISBN 978-3-642-00488-9 e-ISBN 978-3-642-00489-6
DOI 10.1007/978-3-642-00489-6
Springer Dordrecht Heidelberg London New York

Library of Congress Control Number: 2009922261

Cover design: WMXDesign, Heidelberg, Germany

Printed on acid-free paper

Springer is part of Springer Science+Business Media (www.springer.com)

Preface

MicroRNAs (miRNAs), endogenous noncoding regulatory mRNAs of around 22-nucleotides long, have rapidly emerged as one of the key governors of the gene expression regulatory program in cells of varying species. Accumulating evidence suggests that miRNAs constitute a novel, universal mechanism for fine regulation of gene expression in all organisms, "fine tuning" the cellular phenotype during delicate processes. Owing to their ever-increasing number in the mammalian cells and their ever-increasing implication in the control of the fundamental biological processes (such as development, cell growth and differentiation, cell death, etc), miRNAs have now become a research subject capturing major interest of scientists worldwide. Moreover, with recent studies revealing the macro roles of miRNAs in the pathogenesis of adult humans, we have now entered a new era of miRNA research. The exciting findings in this field have inspired us with a premise and a promise that miRNAs will ultimately be taken to heart for the therapy of human disease. Yet these mysterious tiny molecules still remain mystifying in their cellular function and pathological role. While miRNAs have been considered potential therapeutic targets for disease treatment, it remains obscure what strategies we can use to achieve this goal. Thorough understanding of these molecules is obviously a prerequisite for realizing their rousing promise, which calls for an urgent need to develop apt technologies for the purpose.

In the past years, we have witnessed the rapid development of many creative, innovative, inventive techniques and methodologies pertinent to miRNA research and applications. These technologies have convincingly demonstrated their efficacy and reliability in producing gain-of-function or loss-of-function of miRNAs, providing new tools for elucidating miRNA functions and opening up a new avenue for the development of new agents targeting miRNAs for therapeutic aims. These stimulating advances prompted me to propose the concept of microRNA interference (miRNAi): *Manipulating the function, stability, biogenesis or expression of miRNAs to interfere with the expression of their target protein-coding mRNAs to alter the cellular functions.* This new thought motivated me to write this book entitled *MicroRNA Interference Technologies (miRNAi Technologies).*

The aim of this book is to provide comprehensive descriptions of the strategies and methodologies for interfering miRNA expression, biogenesis and function and their applications in miRNA research and new drug design using miRNAs as therapeutic targets. It is my expectation that from this book readers will be able to acquire a basic knowledge of miRNAs and the new concepts pertinent to miRNAi, gain insight into the principles of various miRNAi technologies and master the key steps of miRNAi protocols.

miRNAi Technologies contains 13 chapters. It begins with Chapter 1 on the updated knowledge of miRNA biology and their potential as therapeutic targets for human disease. Chapter 2 introduces four new concepts pertinent to miRNAs, which are of pivotal importance for our understanding and application of the miRNAi technologies. These new concepts are (1) the "miRNA Interference (miRNAi)" concept, (2) the "miRNA as a Regulator of a Cellular Function" concept, (3) the "One-Drug, Multiple-Target" concept and (4) the "miRNA Seed Family" concept. Chapter 2 also gives a laconic introduction of miRNAi strategies and the perspectives of miRNAi technologies in a general term. From Chaps. 3–13, each chapter introduces one of the miRNAi technologies with detailed descriptions of state-of-the-art design, step-by-step directive protocols, principles of action, applications to basic research, R&D and clinical therapy and advantages and limitations of the technologies. Chapters 3–6 describe various gain-of-function miRNAi technologies and chapters 7–13 introduce the loss-of-function miRNAi technologies.

Each chapter also contains illustrations, flowcharts and tables for easier and straightforward understanding of the contents. Though step-by-step protocols are provided for each miRNAi technology, it is not my attempt to give very detailed, problem-proof procedures by including information like compositions of solutions, conditions of reactions and materials. Instead, I intend to provide readers with a guideline for designing and setting up the protocols for their own particular uses. This also leaves room for readers to make their own improvements and innovations of the technologies.

This book is written for: (1) Fundamentalists (starting from graduate students to PI) in the field of studies involving miRNAs, in universities and research institutions; (2) Pharmacologists and gene therapists involving translational studies on drug development; (3) Pharmaceutical companies involving R&D in target searching and drug design and (4) Medical practitioners from residents to professors of various types of medical fields.

Canada Zhiguo Wang

Acknowledgement

Writing a science book on a very specialized subject is absolutely not a solo but an orchestral achievement; though I am the sole author, hundreds of workers whose studies are cited, or are left out by accident, in the reference lists have contributed importantly and enormously to this book by providing their invaluable information and sharing their fascinating findings in their wonderful publications. I wish to give my sincere thanks to all these people, whether I knew them in person or not; I know their names from their papers and their works by heart. Without their sweat and intellect, this book would have been absolutely impossible.

My special thanks go to Dr Stanley Nattel, my former PhD supervisor and boss, who has given me everything I need to start my independent scientific career and to Dr Baofeng Yang, my best friend, with whom I have the pleasure of sharing a stage and working with and whose words over the years has taught me much about myself and the mysterious ways of life, the scientific life.

I wish to give my gratitude to my dear wife Xiaofan Yang who has had to live with my fluctuating attitude as I wrote these chapters. She has stood by me and meanwhile taken care of our kids and the nuts and bolts of our household. She has been my inspiration and support that has driven me onwards. She has kept me focused on everything I have done. Thank you my darling.

Much of what I have learnt over the years came as the result of being a father to two wonderful and delightful children, Ritchie and Jennifer, both of whom, in their own ways inspired me and, subconsciously contributed to this book, despite that they do not (and may never) understand the contents. The joyfulness they give me is definitely another driving force for me to complete the mission-impossible book writing – thanks kids!

I wish to express my appreciation to my wonderful fellows and students, Guorong Chen, Haijun Zhang, Huixian Lin, Huizhen Wang, Jiangchun Zhang, Jiening Xiao and Xiaobin Luo (listed in alphabetic order), for their continuous motivation, intelligence, hard work and understanding. I am also particularly grateful to Dr Jiening Xiao who convinced me to commence the miRNA business. We have been able to openly share knowledge, ideas and numerous tips,all of which culminated in the completion of this book. Of course, we went through some difficult and cheerful times too, sharing together the frustration and exuberance. Without everyone's creative studies, I would not be the author of this book. What an amazing research team! Thank you all, guys!

Contents

1 miRNAs Targeting and Targeting miRNAs 1
1.1 miRNA Biology 2
1.1.1 miRNAs Biogenesis 2
1.1.2 miRNAs Actions 6
1.2 miRNA Expression, Mutation and Polymorphism 8
1.2.1 miRNA Expression 8
1.2.2 miRNA Mutation 9
1.2.3 miRNA Polymorphism 10
1.3 miRNAs and Human Disease 12
1.3.1 miRNAs and Developmental Disorders 12
1.3.2 miRNAs and Apoptosis 17
1.3.3 miRNAs and Cancer 20
1.3.4 miRNAs and Cardiovascular Disease 23
1.3.5 miRNAs and Neuronal Disease 28
1.3.6 miRNAs and Viral Disease 30
1.3.7 miRNAs and Metabolic Disorders 34
1.3.8 miRNA and Epigenetics 36
1.4 miRNAs as Therapeutic Targets 38
1.4.1 Strategies for Therapeutic Modulation of miRNAs 39
1.4.2 Approaches for Therapeutic Modulation of miRNAs 41
References 41

2 miRNA Interference Technologies: An Overview 59
2.1 New Concepts of miRNAi Technologies 59
2.1.1 "miRNAi", A New Concept 59
2.1.2 "miRNA as a Regulator of a Cellular Function", Second New Concept 62
2.1.3 "One-Drug, Multiple-Target", Third New Concept 63
2.1.4 "miRNA Seed Family", Another New Concept 64

2.2 General Introduction to miRNAi Technologies 66
2.2.1 miRNA-Targeting Technologies 66
2.2.2 Targeting-miRNA Technologies 66
2.3 miRNAi Technologies in Basic Research and Drug Design 67
References 71

3 Synthetic Canonical miRNA Technology 75
3.1 Introduction 75
3.2 Protocols 76
3.2.1 Designing SC-miRNAs 76
3.2.2 Validating SC-miRNAs 78
3.3 Principle of Actions 86
3.4 Applications 87
3.5 Advantages and Limitations 88
References 88

4 miRNA Mimic Technology 93
4.1 Introduction 93
4.2 Protocols 94
4.2.1 Designing miR-Mimics 95
4.2.2 Validating miR-Mimics 96
4.3 Principle of Actions 97
4.4 Applications 98
4.5 Advantages and Problems 99
References 100

5 Multi-miRNA Hairpins and Multi-miRNA Mimics Technologies 101
5.1 Introduction 101
5.2 Protocols 102
5.2.1 Construction of Multi-miRNA Hairpins 102
5.2.2 Construction of Multi-miRNA Mimics (Chen et al. 2009) .. 105
5.3 Principle of Actions 107
5.4 Applications 107
5.5 Advantages and Limitations 108
References 109

6 miRNA Transgene Technology 111
6.1 Introduction 111
6.2 Protocols 112
6.2.1 Conventional Transgene Methods 112
6.2.2 Artificial Intronic miRNA Methods 115
6.2.3 Cre-loxP Knock-in Methods 120
6.3 Principle of Actions 120
6.3.1 Conventional Transgene Methods 121
6.3.2 Artificial Intronic miRNA Methods 121

6.4 Applications 122
6.5 Advantages and Limitations 123
References 125

7 Anti-miRNA Antisense Oligonucleotides Technology 127
7.1 Introduction 128
7.2 Protocols 129
7.2.1 Designing AMOs 129
7.2.2 Modifying AMOs 131
7.2.3 Monitoring Delivery Efficiency of AMOs 133
7.2.4 Evaluating Functional Effectiveness of AMOs 134
7.3 Principle of Actions 135
7.4 Applications 136
7.5 Advantages and Limitations 138
References 140

8 Multiple-Target Anti-miRNA Antisense Oligonucleotides Technology 145
8.1 Introduction 145
8.2 Protocols 147
8.2.1 Designing MT-AMOs 147
8.2.2 Validating MT-AMOs 147
8.3 Principle of Actions 148
8.4 Applications 149
8.5 Advantages and Limitations 149
References 150

9 miRNA Sponge Technology 153
9.1 Introduction 153
9.2 Protocols 154
9.2.1 Designing miRNA Sponges 154
9.2.2 Validating miRNA Sponges 156
9.3 Principle of Actions 157
9.4 Applications 157
9.5 Advantages and Problems 158
References 158

10 miRNA-Masking Antisense Oligonucleotides Technology 161
10.1 Introduction 161
10.2 Protocols 162
10.3 Principle of Actions 163
10.4 Applications 164
10.5 Advantages and Limitations 166
References 166

11 Sponge miR-Mask Technology ... 167
11.1 Introduction ... 167
11.2 Protocols ... 168
11.3 Principle of Actions ... 169
11.4 Applications ... 172
11.5 Advantages and Limitations ... 172
References ... 173

12 miRNA Knockout Technology ... 175
12.1 Introduction ... 175
12.2 Protocols ... 176
12.2.1 Homologous Recombination Methods ... 176
12.2.2 Cre-loxP Methods ... 177
12.2.3 FLP-FRT Deletion Methods ... 177
12.3 Principle of Actions ... 179
12.4 Applications ... 180
12.5 Advantages and Limitations ... 181
References ... 182

13 Dicer Inactivation Technology ... 183
13.1 Introduction ... 183
13.2 Protocols ... 184
13.2.1 Neomycin-Expression Cassette Methods (Homologous Recombination) ... 184
13.2.2 Cre-loxP Methods ... 185
13.3 Principle of Actions ... 185
13.4 Applications ... 185
13.5 Advantages and Limitations ... 187
References ... 188

Index ... 191

Chapter 1
miRNAs Targeting and Targeting miRNAs

Abstract With the recent advance of research into microRNAs (miRNAs), this category of endogenous noncoding small ribonucleic acids (19–25 nts in length) has rapidly emerged as one of the central regulators of expression of an extensive repertoire of genes. MiRNAs are an abundant RNA species constituting >2% of the predicted human genes (>1,000 genes), which regulates ~30% of protein-coding genes. Some miRNAs are expressed at >1,000 copies per cell. Thousands of miRNAs have been identified in several organisms including humans, some of which are registered in the miRBase Registry (http://microrna.sanger.ac.uk/registry/; the Wellcome Trust Sanger Institute). Computational prediction suggests an even larger number of miRNAs (~25,000 in humans) exist in mammalian genome that are still to be identified [Miranda KC, Huynh T, Tay Y, Ang YS, Tam WL, Thomson AM, Lim B, Rigoutsos I, Cell 126:1203–1217, 2006; Cummins JM, He Y, Leary RJ, Pagliarini R, Diaz LA Jr, Sjoblom T, Barad O, Bentwich Z, Szafranska AE, Labourier E, Raymond CK, Roberts BS, Juhl H, Kinzler KW, Vogelstein B, Velculescu VE, Proc Natl Acad Sci USA 103:3687–3692, 2006.]

The high sequence conservation across metazoan species suggests strong evolutionary pressure and participation of miRNAs in essential biological processes such as cell proliferation, differentiation, apoptosis, metabolism, stress and so forth [Lewis BP, Shih IH, Jones-Rhoades MW, Bartel DP, Burge CB, Cell 115:787–798, 2003; Lewis BP, Burge CB, Bartel DP, Cell 120:15–20, 2005; Jackson RJ, Standart N, Sci STKE 23:243–249, 2007; Nilsen TW, Trends Genet 23:243–249, 2007; Pillai RS, Bhattacharyya SN, Filipowicz W, Trends Cell Biol 17:18–126, 2005; Alvarez-Garcia I, Miska EA, Development 132:4653–4662, 2005; Ambros V, Nature 431:350–355, 2004.] MiRNAs are also critically involved in a variety of pathological processes including human disease, such as developmental malformations, cancer, cardiovascular disease, neuronal disorders, metabolic disturbance and viral disease. MiRNAs have been considered a part of the epigenetic program in organisms.

The initial discovery of small temporal RNAs (now known as miRNAs) is credited to the pioneer work described by Lee et al. [Cell 75: 843–854, 1993] and Wightman et al. [Cell 75: 855–562, 1993] in their effort to search for a

Z. Wang, *MicroRNA Interference Technologies*,
DOI: 10.1007/978-3-642-00489-6_1, © Springer-Verlag Berlin Heidelberg 2009

protein responsible for the disruption of the timing of larval to adult developmental stages due to the lin-4 mutation in the nematode worm *C. elegans*. They identified a small RNA from the locus, which bound with partial complementarity to the 3′-untranslated region (3′UTR) of lin-14 mRNA and negatively regulated lin-14 expression posttranscriptionally. These studies however did not arouse major attention in the scientific community until the let-7 (lethal-7) mutation, which also resulted in disruption of developmental timing in *C. elegans*, was mapped to another small RNA [Reinhart BJ, Slack FJ, Basson M, Pasquinelli AE, Bettinger JC, Rougvie AE, Horvitz HR, Ruvkun G, Nature 403:901–906, 2000]. From that point, researchers began to realize that the let-7 miRNA sequence, along with its expression during development, was conserved in animals from arthropods to humans [Pasquinelli AE, Reinhart BJ, Slack F, Martindale MQ, Kuroda MI, Maller B, Hayward DC, Ball EE, Degnan B, Muller P, Spring J, Srinivasan A, Fishman M, Finnerty J, Corbo J, Levine M, Leahy P, Davidson E, Ruvkun G, Nature 408:86–89, 2000], indicating that miRNAs represent an ancient mechanism of gene regulation. Thus, lin-4 represents the first founding member of the miRNA family that can downregulate the protein lin-14 and let-7 is the second miRNA mediating translational repression of lin-41. However, the tidal wave of miRNA that hit the field of biology was not stirred up until three hallmark papers were simultaneously published in the journal *Science*, which reported the presence of large numbers of small, noncoding RNAs in *Drosophila*, *Caenorhabditis elegans* and mammalian cells [Lau N, Lim L, Weinstein E, Bartel DP, Science 294:858–862, 2001; Lee RC, Ambros V, Science 294:862–864, 2001; Lagos-Quintana M, Rauhut R, Lendeckel W, Tuschl T, Science 294:853–858, 2001.] Thereafter, this new class of small regulatory RNAs gained its big name, miRNA and began to garner interest of scientists worldwide.

1.1 miRNA Biology

1.1.1 miRNAs Biogenesis

Genes for miRNAs are located in the chromosomes and many of them are identified in clusters that can be transcribed as polycistronic primary transcripts. Some miRNAs are encoded by their own genes and others are encoded by the sequences as a part of the host protein-coding genes. Based on the genomic arrangement of miRNA genes, miRNAs can be grouped into two classes:

1. Intergenic miRNAs (miRNA-coding genes located in-between protein-coding genes),
2. Intragenic miRNAs (miRNA-coding genes located within their host protein-coding genes). Further, the intragenic miRNAs can be divided into the following subclasses:

   (a) Intronic miRNAs (miRNA-coding genes located within introns of their host protein-coding genes),

(b) Exonic miRNAs (miRNA-coding genes located within exons of host protein-coding genes),
(c) 3′UTR miRNAs (miRNA-coding genes located within 3′UTR of host protein-coding genes).
(d) 5′UTR miRNAs (miRNA-coding genes located within 5′UTR of host protein-coding genes).

According to our analysis, for the human miRNAs identified thus far, the majority of miRNAs belong to intergenic and intronic miRNAs being comprised of ~42 and ~44% of the total, respectively and the other three categories are rare with the exonic miRNAs being ~7%, 3′UTR miRNAs being 1.5% and 5′UTR miRNAs being 1%.

Clearly, miRNAs either have their own genes or are associated with their host genes; accordingly, miRNAs are generated by two different mechanisms. Biogenesis of miRNAs can be summarized as a five-step process as detailed below (see also Figs. 1.1 and 1.2).

1. Generation of primary miRNAs: transcription of miRNA genes. The intergenic miRNA genes are first transcribed as long transcripts, called primary miRNAs (pri-miRNAs) mostly by RNA polymerase II or RNA polymerase III (Ying and Lin 2005). The pri-miRNAs are capped and polyadenylated and can reach several

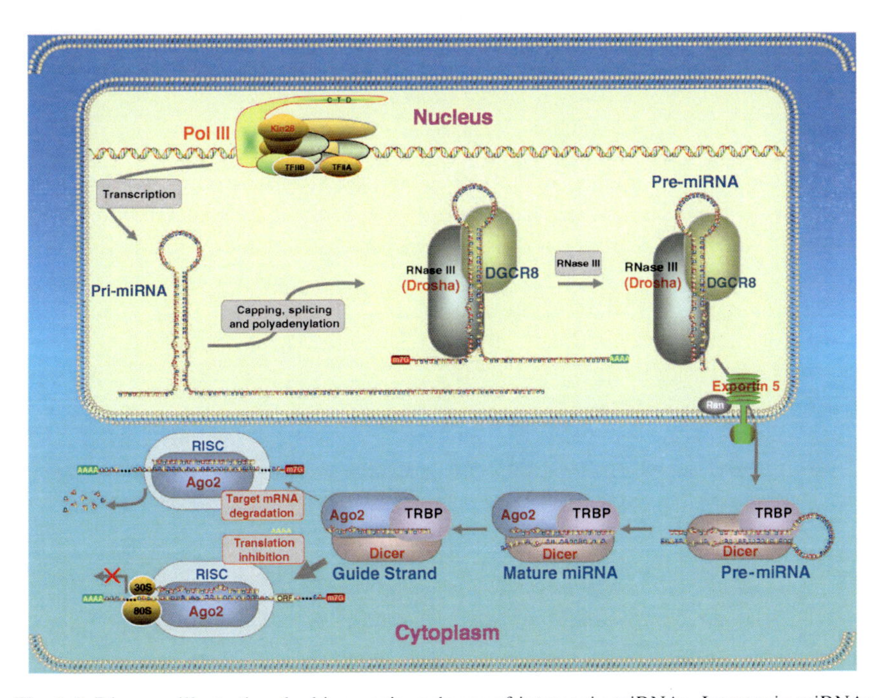

Fig. 1.1 Diagram illustrating the biogenesis pathway of intergenic miRNAs. Intergenic miRNAs have their own genes and their transcription is likely driven by Pol III. Pol III: polymerase III, Ago: Argonaute protein-2

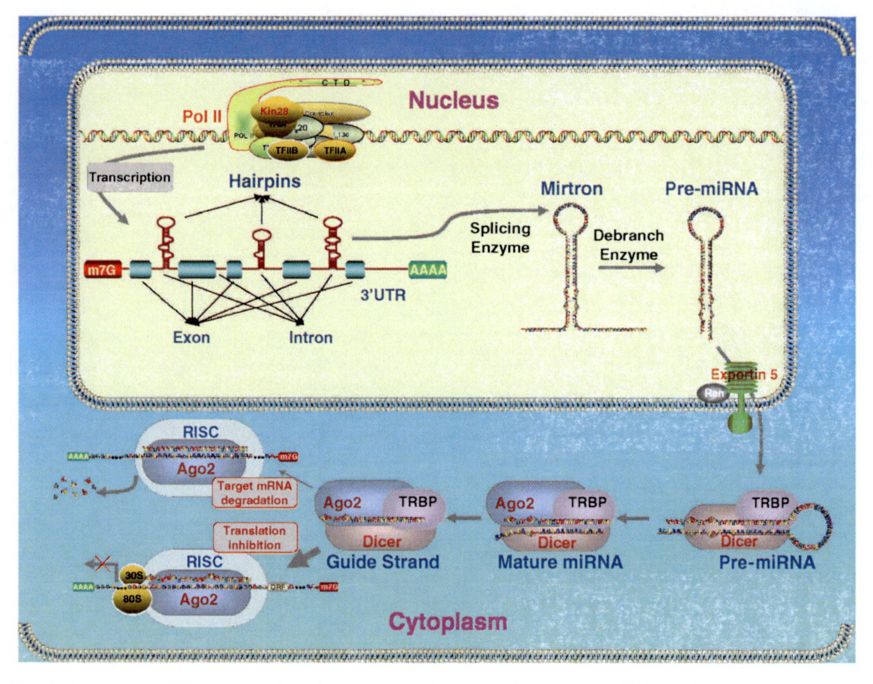

Fig. 1.2 Diagram illustrating the biogenesis pathway of intragenic miRNAs. Intragenic miRNAs are generated by the hairpin structures within host genes (mostly in introns) and they are normally transcripted along with their host genes by Pol II. Pol II: polymerase III, Ago: Argonaute protein-2

kilobases (kb) in length (Cullen 2004; Kim 2005). The clustered miRNA genes in polycistronic transcripts are likely to be coordinately regulated (Bartel 2004). The intronic miRNAs are processed by sharing the same promoter and other regulatory elements of the host genes. They are first transcribed along with their host genes by RNA polymerase II and then processed by Drosha independent pathway from excised introns by the RNA splicing machinery for their biogenesis in *Drosophila*, *C elegans* and mammals (Berezikov et al. 2007; Okamura et al. 2007; Ruby et al. 2007).

2. Generation of precursor miRNAs: endonuclease processing of pri-miRNAs. The pri-miRNAs are processed to precursor miRNAs (pre-miRNAs) by the RNase endonuclease-III Drosha and its partner DGCR8/Pasha in the nucleus (Lee et al. 2002b; Denli et al. 2004; Gregory et al. 2004; Landthaler et al. 2004). These pre-miRNAs are $\sim$60 to $\sim$100 nts with a stem-loop or hairpin secondary structure. Specific RNA cleavage by Drosha predetermines the mature miRNA sequence and provides the substrates for subsequent processing steps. Cleavage of a pri-miRNA by microprocessor begins with DGCR8 recognizing the single-stranded RNA (ssRNA)–double-stranded RNA (dsRNA) junction typical of a pri-miRNA (Han et al. 2006). Then, Drosha is brought close to its substrate through

interaction with DGCR8 and cleaves the stem of a pri-miRNA ∼11 nt away from the two single-stranded segments.

miRNA precursor-containing introns have recently been designated "mirtrons" (Miranda et al. 2006). Mirtrons are derived from certain debranched introns that fold into hairpin structures with 5′ monophosphates and 3′ 2-nt hydroxyl overhangs, which mimic the structural hallmarks of pre-miRNAs and enter the miRNA-processing pathway (Okamura et al. 2007; Ruby et al. 2007). The discovery of mirtrons suggests that any RNA, with a size comparable to a pre-miRNA and all the structural features of a pre-miRNA, can be utilized by the miRNA processing machinery and potentially give rise to a functional miRNA.

3. Nucleus to cytoplasm translocation of pre-miRNAs. Pre-miRNAs are then exported to the cytoplasm from the nucleus through nuclear pores by RanGTP and exportin-5 (Bohnsack et al. 2004; Lund et al. 2004; Yi et al. 2003). After a pre-miRNA is exported to the cytoplasm, RanGTP is hydrolyzed by RanGAP to RanGDP and the pre-miRNA is released from Exp-5.
4. Generation of mature miRNAs: endonuclease processing of pre-miRNAs. In the cytoplasm, pre-miRNAs are further processed by Dicer in animals, which is a highly conserved, cytoplasmic RNase III ribonuclease that chops pre-miRNAs into ∼22 nt duplexes of mature miRNAs containing a guide strand and a passenger strand (miRNA/miRNA*), with 2-nt overhangs at the 3′ termini (Kim 2005). Like other RNase III family proteins, Dicer interacts with double-stranded RNA-binding protein (dsRBP) partners. In mammalian cells, Dicer associates with transactivation-response element RNA-binding protein (TRBP) and protein activator of the interferon-induced protein kinase (PACT) (Chendrimada et al. 2005; Lee et al. 2006). In plants, miRNAs are cleaved into miRNA:miRNA* duplex possibly by Dicer-like enzyme 1 (DCL1) in the nucleus rather than in the cytoplasm (Bartel 2004; Lee et al. 2002a), then the duplex is translocated into the cytoplasm by HASTY, the plant ortholog of exportin 5 (Bartel 2004). The strands of this duplex separate and release mature miRNA of 19–25 nts in length (Bartel 2004; Lee et al. 2002a). Plant miRNAs undergo further modification by methylation at the 3′ end by HEN1 (Yu et al. 2005).
5. Formation of miRISC. Mature miRNAs become integrated into a RNA-induced silencing complex (RISC) to form the miRNA:RISC complex (miRISC). Only one strand of miRNA/miRNA*, the guide strand, is successfully incorporated into RISC, while the other strand, the passenger strand, is eliminated. Strand selection may be determined by the relative thermodynamic stability of two ends of miRNA duplexes (Khvorova et al. 2003; Schwarz et al. 2003). The strand with less stability at the 5′ end is favorably loaded onto RISC, whereas the passenger strand is released or destroyed. miRISC contains several proteins such as Dicer, TRBP, PACT and Gemin3 but the components directly associated with miRNAs are Argonaute proteins (Ago). These proteins contain four domains: the N-terminal, PAZ, middle and Piwi domains. The PAZ domain binds to the 3′ end of guide miRNA, while the other three domains form a unique structure, creating grooves for target mRNA and guide miRNA interactions (Liu et al.

2004; Song et al. 2004; Ma et al. 2005; Parker et al. 2005). In mammalian cells, four Ago proteins have been identified, all of which can bind to endogenous miRNAs (Meister and Tuschl 2004). Despite the sequence similarity among these Ago proteins, only Ago2 exhibits endonuclease activity to slice complementary mRNA sequences between positions 10 and 11 from the 5′- end of guide strand miRNA. Therefore, human Ago2 is a component not only of miRISC but also of siRISC (siRNA-induced silencing complex), a RISC assembled with exogenously introduced siRNA. The roles of various Ago proteins in mammalian RISC are ambiguous but the division of labor among Ago proteins in *Drosophila* is well-defined. *Drosophila* Ago1 and Ago2 have been shown by biochemical and genetic evidence to participate in two separate pathways: Ago1 interacts with miRNA in translational repression, whereas Ago2 associates with siRNA for target cleavage (Carmell et al. 2002; Okamura et al. 2004).

1.1.2 miRNAs Actions

miRNAs exist in double-stranded form (duplex), activate in single-stranded form (simplex) and act in complex form miRISC. Subsequent binding of a miRNA in the miRISC to the 3′ untranslated region (3′UTR) of its target mRNA through a Watson-Crick base-pairing mechanism with its 5′- end 2 to 8 nts exactly complementary to recognition motif within the target. This 5′- end 2 to 8 nt region is termed "seed sequence" or "seed site" as it is critical for miRNA actions (Lewis et al. 2003, 2005). Partial complementarity with the rest of the sequence of a miRNA also plays a role in producing posttranscriptional regulation of gene expression, presumably by stabilizing the miRNA:mRNA interaction. Moreover, the mid and 3′- end regions of a miRNA may also be important for forming miRISC. Studies have shown that in addition to 3′UTR, coding region and 5′UTR can also interact with miRNAs to induce gene silencing (Jopling et al. 2005; Luo et al. 2008; Tay et al. 2008).

In mammalian species, a miRNA can either inhibit translation or induce degradation of its target mRNA or both, depending upon at least the following factors (see Fig. 1.3):

1. The overall degree of complementarity of the binding site,
2. The number of recognition motif corresponding to 5′- end 2 to 8 nts of the miRNA, and
3. The accessibility of the bindings sites (as determined by free energy states) (Jackson and Standart 2007; Nilsen 2007; Pillai et al. 2007).

The greater the degree of complementarity of accessible binding sites, the more likely a miRNA degrades its targeted mRNA. Perfectly complementary targets (Full miRNA:mRNA interaction) are efficiently silenced by the endonucleolytic cleavage activity of some Argonaute proteins (Hutvágner and Zamore 2002; Yekta et al. 2004; Davis et al. 2005) but the vast majority of predicted targets in animals are only partially paired (Partial miRNA:mRNA interaction) (Lewis et al. 2003, 2005;

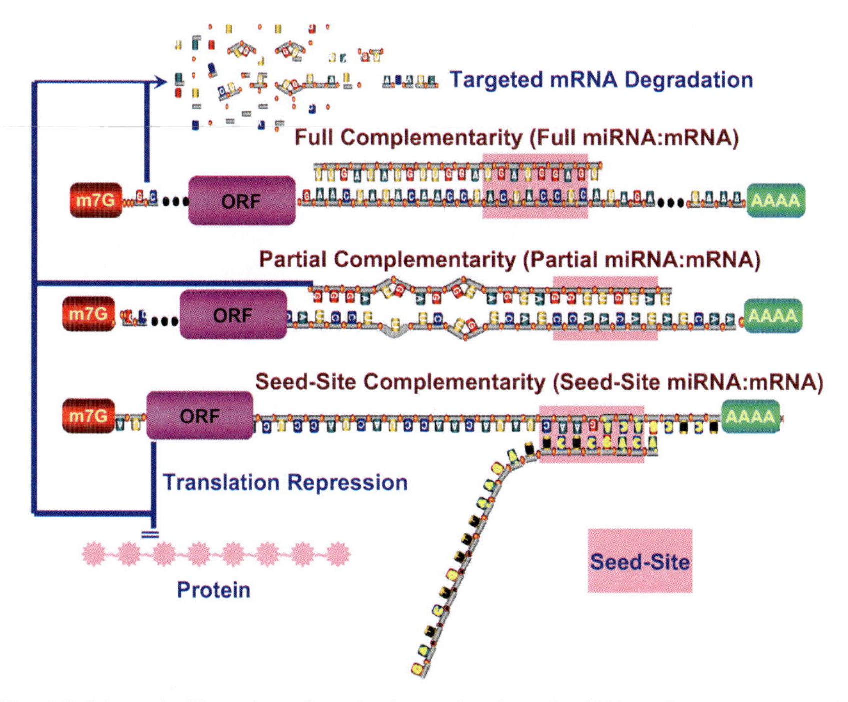

Fig. 1.3 Schematic illustration of mechanisms of action of miRNAs. Full complementarity between a miRNA and its target mRNA (Full miRNA:mRNA) results in targeted mRNA cleavage; Seed-site complementarity (Seed-Site miRNA:mRNA) leads to translation inhibition; and partial complementarity (Partial miRNA:mRNA) gives rise to both targeted mRNA degradation and protein translation repression

Grun et al. 2005; Krek et al. 2005; Rajewsky 2004; Brennecke et al. 2005) and can hardly be cleaved (Haley and Zamore 2004). Some miRNAs have only seed-site complementarity (Seed-Site miRNA:mRNA) and this interaction primarily leads to translation inhibition. And those miRNAs that display imperfect sequence complementarities with target mRNAs primarily result in translational inhibition (Lewis et al. 2003, 2005; Jackson and Standart 2007; Nilsen 2007; Pillai et al. 2007). The mechanisms for translational inhibition remain largely unknown, although inhibition of translation initiation has been identified as one such mechanism by several studies (Humphreys et al. 2005; Pillai et al. 2005). Greater actions may be elicited by a miRNA if it has more than one accessible binding site in its targeted miRNA, presumably by the cooperative miRNA:mRNA interactions from different sites. mRNA degradation by miRISC is initiated by deadenylation and decapping of the targeted mRNAs (Pillai et al. 2007). A recent study demonstrated, however, that miRNAs can also act to enhance translation when AU-rich elements and miRNA target sites coexist at proximity in the target mRNA and when the cells are in the state of cell-cycle arrest (Vasudevan et al. 2007).

In plants, miRNAs base-pair with their mRNA targets by precise or nearly precise complementarity (Wang et al. 2006).

It has been predicted that each single miRNA can have >1,000 target genes and each single protein-coding gene can be regulated by multiple miRNAs (Lewis et al. 2003, 2005; Jackson and Standart 2007; Nilsen 2007; Pillai et al. 2007; Alvarez-Garcia and Miska 2005; Ambros 2004). This is at least partially a result of a lax requirement of complementarity for miRNA::mRNA interaction (Lim et al. 2005). This implies that actions of miRNAs are sequence- or motif-specific but not gene-specific; different genes can have same binding motifs for a given miRNA and a given gene can have multiple binding motifs for distinct miRNAs. *Based on the characteristics of miRNA actions, I postulate that a miRNA should be viewed as a regulator of a cellular function or a cellular program, not of a single gene* (Wang et al. 2008).

1.2 miRNA Expression, Mutation and Polymorphism

1.2.1 miRNA Expression

Expression of some miRNAs is tissue restricted and of others is ubiquitous. The restriction can be qualitative (some miRNAs are expressed exclusively in certain tissue or cell types but not in others) or quantitative (some miRNAs are abundantly expressed only in certain tissue or cell types and modestly in others). For example, miR-122 accounts for 70% of the total miRNA population in the liver. miR-142 and miR-143 constitute ~30% of the total miRNAs in the colon and spleen, respectively (Lagos-Quintana et al. 2002). In the heart, miR-1 accounts for 45% of all murine miRNAs (Lagos-Quintana et al. 2002). The differential tissue distributions of miRNAs suggest tissue – or even cell type – specific functions of these molecules. For instance, the cell lineage-specific miRNA expression patterns may be required to control timing of development and tissue specification (Lagos-Quintana et al. 2002).

To be more appropriate, while each individual miRNA may not be expressed in a tissue/cell-specific manner, the expression profile of miRNAs appears to be tissue/cell-specific. Many miRNAs are enriched in a tissue/cell-specific manner (Landgraf et al. 2007): miR-1, miR-16, miR-27b, miR-30d, miR-126, miR-133, miR-143 and the let-7 family are abundantly but not exclusively expressed in adult cardiac tissue. In addition to cardiomyocytes, the heart contains many other 'noncardiomyocyte' cell types, such as endothelial cells, smooth muscle cells, fibroblasts and immune cells, which may have completely distinct miRNA expression profiles. Indeed (skin) fibroblasts mainly express miR-16, miR-21, miR-22, miR-23a, miR-24, miR-27a and others, an expression pattern that is highly different from that of cardiomyocytes. In artery smooth muscle the most abundant miRNAs are miR-145, let-7, miR-125b, miR-125a, miR-23 and miR-143 (Ji et al. 2007), despite that the "muscle-specific" miR-1 and miR-133 are also expressed in artery smooth muscle. Other miRNAs, such as the let-7 family, miR-126, miR-221 and miR-222, are highly expressed in human endothelial cells (Kuehbacher et al. 2007; Harris et al. 2008). In addition, miRNA expression profiles can change during cardiac development and many miRNAs that are only normally expressed at significant levels in the fetal human heart are re-expressed in cardiac disease, such as heart failure (Landgraf et al. 2007; Bauersachs and Thum, 2007).

Probably more important is the fact that the expression profile of miRNAs is disease-dependent. A particular pathological process may be associated with the expression of a particular group of miRNAs; this is what the signature expression pattern of miRNAs implies. This issue is discussed in the following sections of this chapter.

1.2.2 miRNA Mutation

It is generally believed that the mechanisms that alter the expression of miRNAs are similar to those that change the expression levels of mRNAs of tumor suppressors and oncogenes, i.e., gross genomic aberrations, transcriptional deregulations, epigenetic changes, DNA copy number abnormalities, defects in the miRNA biogenesis machinery and minor mutations affecting the expression level, processing, or target-interaction potential of the miRNA. However, to the best of my knowledge, mutation within the miRNA genes is rare (Diederichs and Haber 2006).

One study (Li et al. 2005) demonstrated that lin-58 alleles contain point mutations in a gene regulatory element of miR-48, a let-7 family member. This mutation causes developmental timing defects in *C. elegans*.

Some miRNA genes are frequently located at fragile sites, as well as in minimal regions of loss of heterozygosity, minimal regions of amplification (minimal amplicons), or common breakpoint regions (Calin et al. 2004b). For example, miR-15a and miR-16-1 genes are located at chromosome 13q14, a region that is frequently deleted in pituitary tumors (Calin et al. 2004b; Bottoni et al. 2005). Deletion mutations in the 3′ flanking region of miR-16-1 transcript have been identified in families with two or more members with chronic lymphocytic leukemia (Calin et al. 2005). Raveche et al. and Colleagues (2007) reported a point mutation in the 3′ flanking sequence of miR-16-1 in a strain of mice prone to autoimmune and B lymphoproliferative disease. They reported decreased levels of expression of miR-16 residing in the mouse D14mit160 region (Raveche et al. 2007). The human region of synteny with mouse D14mit160 is the human 13q.14 where miR-15 and miR-16 reside and a region which, as reported earlier by Calin et al. (2002, 2005), is deleted in chronic lymphocytic leukemia.

1.2.3 miRNA Polymorphism

1.2.3.1 Polymorphism in miRNAs

Sequence variations in miRNA genes (Single nucleotide polymorphism, SNP), including pri-miRNAs, pre-miRNAs and mature miRNAs, could potentially influence the processing and/or target selection of miRNAs.

Duan et al. (2007) reported a systematic survey for miRNA SNPs. They found that SNP is associated with 227 known human miRNAs. Among a total of 323 SNPs, 12 are located within the miRNA precursor and one is at the "seed site" of the mature miR-125a. This miR-125a SNP significantly blocks the processing of pri-miRNA to pre-miRNA, in addition to reducing miRNA-mediated translational suppression.

An example of SNP in pre-miRNAs was documented by Shen et al. (2008), who indentified a G to C SNP located within the sequence of miR-146a precursor, which leads to a change from a G:U pair to a C:U mismatch in its stem region. The predicted miR-146a target genes include BRCA1 and BRCA2, which are key breast and ovarian cancer genes. They found that the variant allele displayed increased production of mature miR-146a from the pre-miRNA compared with the common allele. The binding capacity between the 3′UTR of BRCA1 and miR-146a was found statistically significantly stronger in variant C allele than those in common G allele. The results suggest that breast/ovarian cancer patients with variant C allele miR-146a may have high levels of mature miR-146 and that these variants predispose them to an earlier age of onset of familial breast and ovarian cancers. On the contrary, it was shown that a common G/C SNP within the pre-miR-146a sequence reduced the amount of pre- and mature miR-146a from the C allele 1.9- and 1.8-fold, respectively, compared with the G allele in papillary thyroid carcinoma (Jazdzewski et al. 2008). The reduction in miR-146a led to less efficient inhibition of target genes involved in the Toll-like receptor and cytokine signaling pathway. Association of SNP in the miR-146a gene with the risk for hepatocellular carcinoma has also been reported (Xu et al. 2008).

Hu et al. (2008) conducted a systematic survey of common pre-miRNA SNPs and their surrounding regions and evaluated in detail the association of 4 of these SNPs with the survival of individuals with nonsmall cell lung cancer. They found that a SNP in miR-196a2 was associated with survival in individuals with non-small cell lung cancer; survival was significantly decreased in individuals who were homozygous CC at SNP. In the genotype-phenotype correlation analysis of 23 human lung cancer tissue samples, the SNP was associated with a statistically significant increase in mature miR-196a expression but not with changes in levels of the precursor, suggesting enhanced processing of the pre-miRNA to its mature form. Furthermore, binding assays revealed that the SNP can affect binding of mature miR-196a2-3p to its target mRNA.

Additionally, a SNP in miR-27a genome has been associated with the development of gastric mucosal atrophy in Japanese male subjects (Arisawa et al. 2007).

1.2.3.2 Polymorphism in miRNA Targets

Genetic polymorphisms can reside on miRNA-binding sites. Thus, it is conceivable that the miRNA regulation may be affected by SNPs on the 3′UTRs of target genes. Recently, a novel class of functional SNPs termed miRSNPs/polymorphisms has been reported (Bertino et al. 2007; Chen et al. 2008b; Georges et al. 2007; Landi et al. 2008; Mishra et al. 2008; Yu et al. 2007), defined as a polymorphism present at or near miRNA recognition motifs of target genes that can affect gene expression by interfering with a miRNA function.

Saunders et al. (2007) analyzed publicly available single nucleotide polymorphism (SNP) data in context of miRNAs and their target sites throughout the human genome and found a relatively low level of variation in the functional regions of miRNAs but an appreciable level of variation at target sites. Approximately 400 SNPs have been identified at experimentally verified target sites or predicted target sites that are otherwise evolutionarily conserved across mammals (Saunders et al. 2007). Approximately 250 SNPs potentially create novel target sites for miRNAs in humans. Functional SNPs in 3′UTRs of several genes have been reported to be associated with diseases by affecting gene expression. The mechanism by which these SNPs affect gene expression and induce variability in a cell is that SNPs can either abolish existing binding sites or create illegitimate binding sites.

For instance, a C–T naturally occurring SNP was identified near the miR-24 binding site in the 3′UTR of human dihydrofolate reductase. This SNP leads to a decrease in miR-24 binding resulting in overexpression of its target product and resistance to the enzyme inhibitor methotrexate (Mishra et al. 2007).

It was demonstrated that the myostatin (GDF8) allele of Texel sheep is characterized by a G to A transition in the 3′UTR that creates a target site for miR-1 and miR-206, miRNAs that are highly expressed in skeletal muscle. The result of this mutation is the downregulation of myostatin, which, in turn, contributes to the muscular hypertrophy of Texel sheep (Clop et al. 2006).

The 3′UTR of angiotensin receptor 1 contains an A–C SNP, which has been associated with hypertension in many studies. Sethupathy et al. (2007) found that this SNP results in a relief of repression of angiotensin receptor 1 by miR-155, thereby an elevation of angiotensin receptor 1 levels, which can presumably result in hypertension.

A destabilizing effect of the miRNA::mRNA interaction was described for germline SNPs found in the 3′UTR of the KIT oncogene and corresponding to the target site of miR-221, miR-222 and miR-146 (He et al. 2005a). It is known that thyroid papillary carcinoma is a type of cancer with high familiarity without known genetic bases. It is thus possible that the SNPs in the 3′UTR of KIT mentioned above are abnormalities involved in thyroid cancer predisposition.

Tourette's syndrome, a neurological disorder with an unknown etiology, might result from mistargeting between a miRNA and one of its targets (Abelson et al. 2005). An SNP that correlates with this disease is located in the 3′UTR of the SLITRK1 (SLIT and TRK-like 1) gene; this SNP resides within the binding site for miR-189, rendering SLITRK1 a better interacting target of this miRNA.

These observations indicate that SNPs in miRNA can have a differing effect on gene and protein expression and represent another type of genetic variability that can influence the risk of certain human diseases.

1.3 miRNAs and Human Disease

As evidenced below, miRNAs have been increasingly associated with many diverse diseases. Experimental data linking miRNAs to particular pathological states have been overwhelming. Gain-of-function and/or loss-of-function of specific miRNAs appear to be a key event in the genesis of diseases (Table 1.1).

1.3.1 miRNAs and Developmental Disorders

Tissue development depends on the right spatiotemporal expression of relevant protein-coding genes, which are critically regulated in their expression by miRNAs. Indeed, certain spatiotemporal patterns of expression of particular miRNAs have been implicated in the development of tissues/organs. Cell- and/or tissue-specific modulation of miRNA expression levels has been demonstrated to determine the fine-tuning of specific targets that need to be activated or silenced at a certain stage of development (Farh et al. 2005).

Indeed, the first miRNAs were identified by virtue of their effects on worm development. Lin-4 was found accumulating during the first and second larval stages and triggered passage to the third larval stage by repressing the translation of at least two genes, lin-14 and lin-28 (Lee et al. 1993; Wightman et al. 1993). The activity of lin-4 depends on the partial homology of the miRNA to specific regions of the 3′UTRs of the lin-14 and lin-28 mRNAs. A second miRNA, let-7 accumulates during *C.elegans* larval development and triggers passage from late larval to adult cell fates by repressing lin-41 (Reinhart et al. 2000; Pasquinelli et al. 2000). Relatively few miRNAs (e.g., let-7b, miR-130b and miR-367) are detectable during early embryogenesis in the chick epiblast (Darnell et al. 2006). Similarly, in early zebrafish, miR-430 is the only miRNA expressed at an appreciable level (Giraldez et al. 2005).

Mutations in Dicer or Argonaute cause severe developmental phenotypes in several organisms (Grishok et al. 2001; Knight and Bass 2001; Wienholds et al. 2003; Bernstein et al. 2003; Giraldez et al. 2005; Hatfield et al. 2005; Deshpande et al. 2005; Cox et al. 1998). In mice, a Dicer knockout is lethal at day E7.5, pointing to early and essential roles for miRNAs in development and the inability of Dicer null embryonic stem cells to proliferate suggests a role in stem cell maintenance (Bernstein et al. 2003). Block of miRNA biogenesis by tissue-specific deletion of Dicer causes lethality of embryos due to defects in cardiogenesis. When Dicer was efficiently deleted in the heart, death of the embryos by E12.5 due to

Table 1.1 miRNAs known to be involved in the disorders of biological and physiological processes

miRNA	Description	Reference
Developmental disorders		
Lin-4	Key for passage to the third larval stage of *C.elegans* by repressing the translation of at least two genes, lin-14 and lin-28	Lee et al. (1993); Wightman et al. (1993)
let-7	Key for passage from late larval to adult cell fates of *C.elegans* by repressing lin-41	Reinhart et al. (2000); Pasquinelli et al. (2000)
miR-1	Promote myogenesis in *Drosophila* and mouse by targeting histone deacetylase 4 (HDAC4), Hand2	Chen et al. (2006); Kloosterman (2006); Zhao et al. (2005); Kwon et al. (2005); Zhao et al. (2007); Niu et al. (2007)
miR-133	Enhance myoblast proliferation by repressing serum response factor (SRF)	Chen et al. (2006)
Apoptosis		
miR-17-5p, miR-20a, miR-21, miR-133, miR-146a, miR-146b, miR-191, miR-14, bantam, miR-1d, miR-7, miR-148, miR-204, miR-210, miR-216, miR-296, miR-Lat	Anti-apoptotic miRNAs	Ji et al. (2007); Chan et al. (2005); Corsten et al. (2007); Si et al. (2007); Zhu et al. (2007); Gupta et al. (2006); Xu et al. (2007)
let-7 family, miR-15a, miR-16-1, miR-29, miR-34a, miR-34b, miR-34c, miR-1 and miR-214	Pro-apoptotic miRNAs	Cimmino et al. (2005); Raver-Shapira et al. (2007); Tarasov et al. (2007); Tazawa et al. (2007); Welch et al. (2007); Bommer et al. (2007); Xu et al. (2007)
miR-21, miR-24	Anti-apoptotic and pro-apoptotic, depending on cell context	Chan et al. (2005); Si et al. (2007); Cheng et al. (2005)
Cancer		
miR-15/miR-16	Mantle cell lymphoma and B cell chronic lymphocytic leukemia (B-CLL)	Calin et al. (2002); Calin et al. (2004b)
miR-125b, miR-145, miR-21, miR155	Breast cancer	Iorio et al. (2005)

(*continued*)

Table 1.1 (Continued)

miRNA	Description	Reference
Let-7 miR-17-92 cluster	Lung cancer	Takamizawa et al. (2004); Johnson et al. (2005); Calin and Croce (2006); Pasquinelli et al. (2000); Tanzer and Stadler (2004); Hayashita et al. (2005)
miR-221, miR-21, miR-181	Brain cancer	Ciafre et al. (2005); Chan et al. (2005)
miR-143 miR-145	Colon cancer	Michael et al. (2003)
miR-155	Lymphoma	Haasch et al. (2002); Tam (2001); Eis et al. (2005)
Cardiovascular disease		
miR-195, miR-1, miR-133, miR-208, miR-21, miR-18b	Cardiac hypertrophy, Heart failure	van Rooij et al. (2006); van Rooij et al. (2007); Sayed et al. (2007); Carè et al. (2007); Thum et al. (2007); Cheng et al. (2007); Tatsuguchi et al. (2007)
miR-1, miR-133, miR-328	Cardiac arrhythmias	Yang et al. (2007); Luo et al. (2007); Xiao et al. (2007); Zhao et al. (2007); Luo et al. (2008); Zhang et al. (2009)
miR-21, miR-221, miR-222, let-7f, miR-27b	Vascular angiogenesis	Ji et al. (2007); Suárez et al. (2007); Kuehbacher et al. (2007)
Neuronal disease		
miR-124a, miR-128, miR-23, miR-26 and miR-29, miR-9, miR-134	Neuronal development	Krichevsky et al. (2003); Schratt et al. (2006); Miska et al. (2004)
Dicer (no particular miRNAs are yet known)	Fragile X syndrome	Caudy et al. (2002); Jin et al. (2004); Singh (2007)
miR-189	Tourette's syndrome	Abelson et al. (2005)
Viral miRNAs		
miR-H1	Herpesvirus/Simplexvirus (virus-encoded, human host)	Cui et al. (2006a)

(*continued*)

Table 1.1 (Continued)

miRNA	Description	Reference
miR-UL22A, miR-UL36, miR-UL70, miR-UL112, miR-UL148d, miR-US4, miR-US5-1, miR-US5-2, miR-US25-1, miR-US25-2, miR-US33	Herpesvirus/Cytomegalovirus (virus-encoded, human host)	Dunn et al. (2005); Grey et al. (2005); Landgraf et al. (2007); Pfeffer et al. (2005)
miR-BART1 to miR-BART-20, miR-BHRF1-1-, miR-BHRF1-2, miR-BHRF1-3,	Herpesvirus/Lymphocryptovirus (virus-encoded, human host)	Cai et al. (2006); Grundhoff et al. (2006); Pfeffer et al. (2004);
miR-K12-1 to miR-K12-12	Herpesvirus/Rhadinovirus (virus-encoded, human host)	Cai et al. (2006); Grundhoff et al. (2006); Pfeffer et al. (2004); Samols et al. (2005)
Unnamed	Adenovirus (virus-encoded, human host)	Andersson et al. (2005); Aparicio et al. (2006); Sano et al. (2006)
miR-17/92 cluster	Downregulated following HIV infection	Kumar (2007)
miR-122	HCV-1 RNA replication	Esau et al. (2006); Krützfeldt et al. (2005); Jopling et al. (2005)
miR-32	Limit the replication of the retrovirus primate foamy virus	Lecellier et al. (2005)
miR-24, miR-93	Downregulate viral mRNAs of vesicular stomatitis virus	Pedersen et al. (2007)
miR-196, miR-296, miR-351, miR-431, miR-448	Limit HCV RNA accumulation	Pedersen et al. (2007)
Metabolic disorders		
miR-375	Decreases insulin secretion	Poy et al. (2007)
miR-9	Reduction in exocytosis in pancreas (insulin secreting beta cells)	Plaisance et al. (2006)
miR-122	Increase plasma cholesterol levels	Krützfeldt et al. (2005)
miR-9, miR-143	Increase triglyceride accumulation	Esau et al. (2004)

cardiac failure was consistently observed (Zhao et al. 2007). Dicer-deficient embryos developed cardiac failure due to pericardial edema and underdevelopment of the ventricular myocardium. These phenotypes are in agreement with the defects of heart development in zebrafish embryos devoid of Dicer function (Giraldez et al. 2005).

Heart malformations occur in as high as 1% of newborns, presenting a significant clinical problem in our modern world. The role of miRNAs in cardiac development and related disorders has been intensively studied. It is now commonly recognized that miRNAs are a critical player in orchestrating morphogenesis as well as in early embryonic patterning processes. The miRNAs that are significantly upregulated during the differentiation of embryonic stem cells to cardiomyocytes are miR-1, miR-18, miR-20, miR-23b, miR-24, miR-26a, miR-30c, miR-133, miR-143, miR-182, miR-183, miR-200a/b, miR-292-3p, miR-293, miR-295 and miR-335 in mice (Srivastava et al. 1997) and miR-1, miR-20, miR-21, miR-26a, miR-92, miR-127, miR-129, miR-130a, miR-199b, miR-200a, miR-335 and miR-424 in humans (Ivey et al. 2008). A considerable number of miRNAs upregulated in differentiating cardiomyocytes are also enriched in human fetal heart tissue (Thum et al. 2007). The number of detectable miRNAs increases rapidly in tissues derived from all three germ layers (endoderm, ectoderm and mesoderm). An analysis of the miRNAs expressed in undifferentiated mouse embryonic stem cells and differentiating cardiomyocytes was recently published (Lakshmipathy et al. 2007). Among miRNAs detected in early chick heart development, as well as in myotomal skeletal muscles are miR-1 and miR-133 (Kloosterman 2006). These muscle-specific miRNAs determine skeletal muscle proliferation and differentiation (Chen et al. 2006). miR-1 promotes myogenesis by targeting histone deacetylase 4 (HDAC4), a transcriptional repressor of muscle gene expression, whereas miR-133 enhances myoblast proliferation by repressing the serum response factor (SRF), which in turn is required for the transcriptional activation of various other miRNAs.

Moreover, the muscle-specific miRNAs are expressed in the early stages during cardiogenesis. Microarray analysis revealed an increased expression of miR-1 and miR-133 in developing mouse hearts from embryonic day E12.5 through to at least E18.5, indicating a requirement of these miRNAs for cardiac development (Chen et al. 2006). Several lines of evidence have been obtained in support of this notion (Zhao et al. 2005, 2007; Kwon et al. 2005; Niu et al. 2007). Loss-of function of miR-1 in *Drosophila* resulted in embryonic/larval lethality with most of the mutant flies displaying altered sarcomeric gene expression and, in a subset of embryos, an increased number of undifferentiated muscle progenitors (Kwon et al. 2005). In mice, miR-1 is responsible for the inhibition of cardiomyocyte progenitor proliferation via inhibition of translation of Hand2 (Zhao et al. 2005), a transcription factor known to regulate ventricular cardiomyocyte expansion (Srivastava et al. 1997). Many of the embryos from miR-1-2 knockout mice demonstrated ventricular septal defects in a subset that suffer early lethality. The adult miR-1-2-deficient mice had thickened chamber walls attributable to hyperplasia of the heart (Zhao et al. 2007). In contrast, miR-1 gain-of-function led to 100% embryonic fly lethality because of disrupted patterning of cardiac and skeletal muscle with insufficient numbers of cardioblasts. miR-1 controls heart development in mice by regulation of the cardiac transcription factor Hand2 (Zhao et al. 2005). Overexpression of miR-1 in a transgenic mouse model resulted in a phenotype characterized by thin-walled ventricles, attributable to premature differentiation and early withdrawal of cardiomyocytes from the cell cycle (Zhao et al. 2007). Altogether, these results reveal a tight control of spatiotemporal miR-1 expression for proper cardiac and/or skeletal muscle development.

1.3.2 miRNAs and Apoptosis

Apoptosis is an active process that leads to cell death. Unlike necrosis, apoptosis is a complex endogenous gene-controlled event that requires an exogenous signal – stimulated or inhibited by a variety of regulatory factors, such as formation of oxygen free radicals, ischemia, hypoxia, reduced intracellular K^+ concentration and generation of nitric oxide. Apoptosis has been implicated in a variety of human diseases such as heart disease, Alzheimer's disease, cancer, etc. In addition to well-established features of apoptotic cells, such as cell shrinkage, membrane blebbing, chromatin condensation and DNA fragmentation, we have recently identified three intrinsic properties of apoptosis. The first of these is what we term Apoptotic Preconditioning: a brief exposure of cells to a sublethal dose of apoptotic inducers increases the tolerance of the cells to subsequent lethal apoptotic insults (Han et al. 2001). It is a powerful means of cytoprotection from various environmental and intracellular stresses. The second property is Apoptotic Remodeling: a brief exposure of cells to a lethal dose of apoptosis inducers can trigger a signaling transduction cascade that commits cells to death even though the inducers are withdrawn (Han et al. 2004). This property of apoptosis may contribute to anatomical remodeling occurring under several disease conditions, such as the transition from cardiac hypertrophy to heart failure and Alzheimer's disease, as well as to morphogenesis of the organism. Apoptotic Resistance is the third intrinsic property: cells surviving super-lethal apoptotic insults lose their sensitivity to apoptotic inducers (Lu et al. 2008). This property may contribute to the resistance of cells, such as tumor cells, to apoptotic inducers.

To date, no less than 30 individual miRNAs are known to regulate apoptosis. These include the let-7 family, miR-1, miR-1d, miR-7, miR-14, miR-15a, miR-16-1, miR-17 cluster (miR-17-5p, miR-18, miR-19a, miR-19b, miR-20 and miR-92), miR-21, miR-29, miR-34a, miR-133, miR-146a, miR-146b, miR-148, miR-191, miR-204, miR-210, miR-214, miR-216, miR-278, miR-296, miR-335, miR-Lat and bantam. The list is expected to expand quickly with more studies. Indeed, we have performed a bioinformatics prediction of the vertebrate miRNAs available to date in miRBase using a target scan computational analysis with miRBase and miRanda and surprisingly found that nearly all known vertebrate miRNAs ($\sim$93%) have at least one target gene related to cell death and survival or cell cycle and proliferation. This suggests that a majority of, if not all, vertebrate miRNAs can regulate apoptosis in at least some cell types.

Among the known apoptosis-regulating miRNAs, some are designated anti-apoptotic and others proapoptotic miRNAs. This distinction is primarily based upon experimental results from a particular cell type.

1.3.2.1 Anti-apoptotic miRNAs

In general, the following miRNAs are considered anti-apoptotic: miR-17-5p, miR-20a, miR-21, miR-133, miR-146a, miR-146b, miR-191, miR-14, bantam, miR-1d, miR-7, miR-148, miR-204, miR-210, miR-216, miR-296 and miR-Lat. For example, inhibition of miR-17-5p and miR-20a with their antisense oligonucleotides (AMOs) induces apoptosis in lung cancer cells, indicating that miR-17-5p and miR-20a can protect these cells against apoptosis (Ji et al. 2007). Knockdown of miR-21 in cultured glioblastoma cells resulted in a significant drop in cell number. This reduction was accompanied by increases in caspase-3 and 7 enzymatic activities and TUNEL staining (Chan et al. 2005; Corsten et al. 2007). Similarly, in MCF-7 human breast cancer cells, miR-21 also elicits anti-apoptotic effects (Si et al. 2007; Zhu et al. 2007). miR-Lat was reported to protect neuroblastoma cells against apoptotic death (Gupta et al. 2006).

1.3.2.2 Proapoptotic miRNAs

On the other hand, several miRNAs have been referred to proapoptotic ones, which include let-7 family, miR-15a, miR-16-1, miR-29, miR-34a, miR-34b, miR-34c, miR-1 and miR-214. The best example of proapoptotic miRNAs is probably miR-15a and miR-16-1 that when forced to express induces apoptosis in chronic lymphocytic leukemia cells (Cimmino et al. 2005). Equally interesting is the finding that when Wi38 human diploid fibroblasts transduced with an AMO against miR-34 were treated with the apoptosis-inducing agent staurosporine, fewer early apoptotic cells and more viable cells were observed. The authors proposed that miR-34 can mediate key effects associated with p53 function for p53 transactivates gene expression of miR-34 (Raver-Shapira et al. 2007). Similar relationships between miR-34a and p53 have also been confirmed in other cells, such as H1299 human lung cancer cells, MCF-7 human breast cancer cells, U-2OS osteosarcoma cells (Tarasov et al. 2007) and in p53 wild-type HCT116 colon cancer cells (Tazawa et al. 2007). Expression of miR-34a causes dramatic reprogramming of gene expression and promotes apoptosis (Chang et al. 2007). A most recent study further revealed that overexpression of miR-34a causes a dramatic reduction in cell proliferation through the induction of a caspase-dependent apoptotic pathway in three neuroblastoma cell lines – Kelly, NGP and SK-N-AS (Welch et al. 2007). On the other hand, Bommer et al. (2007) demonstrated that the expression of two, miR-34b and miR-34c, is dramatically reduced in 6 of 14 (43%) nonsmall cell lung cancers

(NSCLCs) and that the restoration of miR-34 expression inhibits growth of NSCLC cells.

Previous work on miRNAs and apoptosis has been mostly limited to the context of cancer, while studies on apoptosis regulation by miRNAs in noncancer cells have been sparse. We have recently found that miR-133, one of the muscle-specific miRNAs, produced antiapoptotic actions in neonatal rat ventricular myocytes (Xu et al. 2007). Noticeably, this cytoprotective effect may be weakened in the hypertrophic heart since the miR-133 level has been found significantly reduced under such conditions and this may contribute to increased tendency of apoptosis induction in hypertrophic myocytes (Izumiya et al. 2003; Tea et al. 1999; Teiger et al. 1996). Our study demonstrated that miR-1, another muscle-specific miRNA, promotes apoptosis induced by oxidative stress in cardiac cells counteracting with miR-133 (Xu et al. 2007). Intriguingly, our earlier study revealed that miR-1 level is elevated by 2–3 folds in ischemic myocardium (Yang et al. 2007). Whether this altered miR-1 expression is linked to increased apoptotic cell death in myocardial infarction (Rodríguez et al. 2002; Zidar et al. 2007) merits further study. A more recent study by Yu et al. (2008) clearly demonstrated that miR-1 mediates the apoptotic action of glucose via targeting IGF-1 in H9c2 rat ventricular myocytes. The finding, according to the authors, provides an alternative explanation for the deleterious effects of glucose in the development of cardiomyocyte death and diabetic complications.

1.3.2.3 miRNAs with Both Antiapoptotic and Proapoptotic Properties

Based on the above information, researchers tend to categorize the apoptosis-regulating miRNAs into either an anti-apoptotic or proapoptotic group. Nonetheless, precaution must be taken when attempting to categorize a miRNA by its role in apoptosis. Each miRNA has the potential to regulate >1,000 protein-coding genes that could well be a mixture of some antiapoptotic and proapoptotic genes (Miranda et al. 2006). Whether a miRNA is antiapoptotic or proapoptotic may largely depend upon the cell-specific expression of genes involved in apoptosis and survival. There indeed have been a few instances reinforcing the needs to consider cell context as an important index for the role of miRNAs in apoptosis. As already mentioned above, miR-21 has been considered as antiapoptotic in glioblastoma (Chan et al. 2005) and MCF-7 cells (Si et al. 2007). In HeLa cells, however, miR-21 does the opposite; inhibition of miR-21 increased the number of surviving cells (Cheng et al. 2005). Moreover, inhibition of miR-21 in A549 human lung cancer cells fails to alter cell death or growth (Cheng et al. 2005). Evidently, a same miRNA can have three different actions, antiapoptotic, proapoptotic or neutral, in different cell types. Similarly, miR-24 promotes growth in A549 cells but inhibits growth in HeLa cells (Cheng et al. 2005).

The computational study reported by Wang et al. and Colleagues (2006) presented a global analysis of the interactions between miRNAs and a human cellular signaling network. The authors found that miRNAs predominantly target downstream signaling network components such as transcription factors but less frequently target the genes involved in basic cellular functions and the most upstream signaling network components such as ligands and cell surface receptors. This finding seems to agree with the studies described above.

1.3.3 miRNAs and Cancer

Cancer is a consequence of disordered genome function and has five major stages (initiation, promotion, malignant conversion, progression and metastasis). Different stages have different gene expression profiles, which basically involve balance between two major factors: oncogenes that promote cell proliferation and tumorigenesis and tumor suppressor genes that repress cell division and tumor formation. As miRNAs affect gene expression they were good candidates for keeping the balance between tumor suppressors and oncogenes. It is estimated that more than 200 miRNA sequences discovered in the human genome contribute to the development of cancer (Calin and Croce 2006). Indeed, mounting evidence from both basic and clinical studies indicates that miRNAs are aberrantly expressed in cancer and different types of cancer have different expression profiles of miRNAs, which may ultimately lead to a novel cancer-specific and cancer type-selective treatment strategy. miRNAs have recently been regarded as either oncomiRs or tumor suppressors in carcinogenesis (Dalmay 2008; Hammond 2006; Jay et al. 2007).

Many independent studies on different tissues demonstrated that cancer cells have different miRNA profiles compared with normal cells suggesting that miRNA profiles can be used for diagnosis. The power of miRNA profiling in cancer diagnosis has been established in various types of cancer such as chronic lymphocytic leukemia (Calin et al. 2004a), breast (Iorio et al. 2005), lung cancers (Yanaihara et al. 2006) and six solid tumors including breast, colon, lung, pancreas, prostate and stomach (Volinia et al. 2006). The superiority of miRNA profiling versus mRNA expression for diagnosing and classifying human cancers has been well noticed (Lu et al. 2005), despite that a miRNA profile contains much less information than an mRNA profile (a few hundreds versus 35,000).

1.3.3.1 miRNAs and Leukemia

The first evidence that miRNA expression could be altered in cancer came from the observation that the miR-15/miR-16 gene cluster is located in a genomic region commonly deleted in chronic lymphocytic leukemia (CLL) and that their expression is frequently downregulated in CLL (Calin et al. 2002). A majority of DNA alterations occur in a region on chromosome 13 that are associated with mantle cell lymphoma and B cell chronic lymphocytic leukemia (B-CLL) (Calin et al. 2004b). Interestingly, miRNAs miR-15a and miR-16-1 reside between exons 2 and 5 of the LEU2 gene, a tumor suppressor gene, in human chromosome 13. LEU2 has been excluded. Furthermore, miR-15a and miR-16-1 are found to be downregulated or deleted in 70% of tumor cells from patients with B-CLL.

1.3.3.2 miRNAs and Breast Cancer

Breast cancer is the most common cancer amongst females. miRNA profile of clinical samples of 76 neoplastic and 10 normal breast tissues was established using microarrays containing probes against all known miRNAs. Several miRNAs (miR-125b, miR-145, miR-21 and miR155) showed lower levels in cancer samples (Iorio et al. 2005). The same study also found that the downregulation of these miRNAs correlated with tumor stage, estrogen and progesterone receptor expression, proliferation index and vascular invasion.

1.3.3.3 miRNAs and Lung Cancer

One of the most common cancers of adults in economically developed countries is lung cancer. Reduced expression of let-7 showed good correlation with shortened postoperative survival (Takamizawa et al. 2004). In vitro experiments confirmed this observation; increased expression of let-7 in lung adenocarcinoma cell lines suppresses cell proliferation. Tumor suppressor activity of let-7 was elucidated by validating two predicted target genes. RAS and MYC, key oncogenes in lung cancer, contain several let-7 target sites and are directly regulated by let-7 (Johnson et al. 2005). Let-7 is less expressed in human lung cancers (Calin and Croce 2006). Lower survival rate is observed in patients with diminished expression of let-7. Overexpression of let-7 inhibits growth of lung cancer cells in vitro (Pasquinelli et al. 2000).

On the other hand, the miRNA cluster miR-17-92 contains 14 similar miRNAs and is present in three copies in the human genome (Tanzer and Stadler 2004). The expression of this cluster is strongly upregulated in lung cancer, especially in small-cell lung cancer, which is the most aggressive form of lung cancer (Hayashita et al. 2005). In line with the oncogene activity of these miRNAs their predicted targets are PTEN and RB2, two known tumor suppressor genes (Lewis et al., 2003).

1.3.3.4 miRNAs and Brain Cancer

Human brain cancer is one of the most difficult clinical problems. Gliobastoma multiform is the most common form of brain cancer in human but its development is poorly understood. miR-221 showed higher whilst miR-181 showed lower levels in gliobastoma samples than in normal tissues (Ciafre et al. 2005). Chan et al. (2005) found miR-21 upregulated in gliobastoma and experimental suppression of miR-21 activity led to increased caspase dependent cell death in cultured gliobastoma cells.

1.3.3.5 miRNAs and Colon Cancer

miRNAs were identified in colon tissue and two of them (miR-143 and miR-145) showed reduced expression in different stages of colorectal neoplasia (Michael et al. 2003). Interestingly, the primary transcripts of these two miRNAs were not changed in cancer samples, suggesting that the altered level of mature miRNAs is due to differential processing in normal and cancer tissues.

1.3.3.6 miRNAs and Lymphoma

The BIC gene has been linked to several types of lymphomas (cancer of the lymphocytes, a type of white blood cell in the vertebrate immune system). However, the BIC gene does not encode for a protein and it is considered as a noncoding RNA with oncogene activity. BIC expression is usually higher in Hodgkin and Burkitt lymphoma than in normal lymphoid cells (Haasch et al. 2002) but the mechanism of BIC-induced lymphoma is poorly understood. Comparison of BIC sequences of several species revealed that the conserved region contains a miRNA gene (miR-155) (Tam 2001). These observations inspired a study of miR-155 accumulation in B-cell-derived lymphomas, which found a consistently high level of miR-155 expression (Eis et al. 2005). A putative target gene of miR-155 is the transcription factor PU1 that is necessary for B-cell differentiation.

Several important points concerning miRNAs and cancer are raised from the above findings.

1. Deregulated expression of miRNAs has been characterized in a plethora of human cancers (Calin et al. 2004a, 2005; Iorio et al. 2005; Yanaihara et al. 2006; Volinia et al. 2006). The miRNAs that have attracted tremendous attention from both basic scientists and clinicians include miR-21, miR-34a, miR-34b, miR-34c, miR-15a, miR-16-1, miR-17-5p, miR-20a and let-7.
2. Some of the miRNAs are conserved across cancer types and others are more specifically expressed in certain types of human cancer. For example, miR-21, miR-191 and miR-17-5p are abnormally overexpressed in six different solid tumors, including breast, colon, lung, pancreas, prostate and stomach (Volinia et al. 2006). By comparison, other miRNAs such as miR-155, miR-146 and miR-20a are more restricted to certain types of tumors.
3. Studies have demonstrated the feasibility of targeting relevant miRNAs to facilitate cancer cell death and/or to inhibit cancer cell growth under in vitro conditions. Use of the xenograft mouse model to confirm anti-tumor efficacy under in vivo conditions has also been documented.

Si et al. (2007) showed that one transient transfection producing knockdown of miR-21 by AMO is sufficient to cause substantial inhibition of tumor growth in MCF-7 human breast cancer cells and in the xenograft tumors generated by MCF-7 cells, by effectuating an increase in apoptosis associated with downregulation of

Bcl-2 expression. This finding with miR-21 and MCF-7 in nude mice was confirmed by another group (Zhu et al. 2007).

Intracranial graft survival of gliomas following miR-21–knockdown (by AMO) was diminished, as indicated by a sharp reduction of tumor volume in vivo compared with control–treated gliomas (Corsten et al. 2007).

The study reported by Meng et al. (2007) described the anticancer potential of let-7 depletion by AMO and the underlying molecular mechanisms. The authors demonstrated an increase in basal expression of phosphor-Stat3 and decrease in NF2 in homogenates from xenograft tumors generated by Mz-IL-6 human malignant cholangiocytes injected subcutaneously. Intratumoral administration of let-7a AMO increased NF2 and decreased phosphor-Stat3 expression in Mz-IL-6 xenografts in vivo. Moreover, they observed a decrease in tumor growth consistent with increased gemcitabine toxicity in response to let-7a AMO when compared with tumors that were untreated.

Subcutaneous administration of exogenous miR-34a considerably suppressed in vivo growth of HCT116 and RKO human colon cancer cells in tumors in nude mice (Tazawa et al. 2007). The c13orf25/miR-17 cluster, which is responsible for 13q31-q32 amplification in malignant lymphoma, contains the miRNA-17-18-19-20-92 polycistron. Tazawa et al. (2007) demonstrated that nude mice injected with rat fibroblasts transfected with both miR-17 polycistron and Myc showed accelerated tumor growth compared to those injected with Myc-only transfected cells. The finding suggests an anti-miR-17 antisense should be able to suppress tumor growth.

1.3.4 miRNAs and Cardiovascular Disease

Cardiovascular disease is the leading cause of morbidity and mortality in developed countries. The pathological process of the heart is associated with an altered expression profile of genes that are important for cardiac function. The implications of miRNAs in the pathological process of the cardiovascular system have only recently been appreciated but the research on miRNAs in relation to cardiovascular disease has now become a rapidly evolving field.

1.3.4.1 miRNAs and Cardiac Hypertrophy and Heart Failure

In response to stress (such as hemodynamic alterations associated with myocardial infarction, hypertension, aortic stenosis, valvular dysfunction, etc.), the adult heart undergoes a remodeling process and hypertrophic growth to adapt to altered workloads and to compensate for the impaired cardiac function. Hypertrophic growth manifests enlargement of cardiomyocyte size and enhancement of protein synthesis through the activation of intracellular signaling pathways and transcriptional mediators in cardiac myocytes. The process is characterized by a reprogramming of cardiac gene expression and the activation of 'fetal' cardiac genes (McKinsey and Olson 2005). Recent studies revealed an important role for specific miRNAs in the control of hypertrophic growth and chamber remodeling of the heart and point to miRNAs as potential therapeutic targets in heart disease.

The first common finding is that an array of miRNAs is significantly altered in their expression, some up- and some downregulated. The second common finding is that single miRNAs can critically determine the progression of cardiac hypertrophy.

For example, Olson's group reported >12 miRNAs that are up- or downregulated in cardiac tissue from mice in response to transverse aortic constriction (TAC) or expression of activated calcineurin, stimuli that induce pathological cardiac remodeling (van Rooij et al. 2006). Many of these miRNAs were found similarly regulated in failing human hearts. Forced overexpression of stress-inducible miRNAs induced hypertrophy in cultured cardiomyocytes. Particularly, overexpression of miR-195 alone, which is upregulated during cardiac hypertrophy, is sufficient to induce pathological cardiac growth and heart failure in transgenic mice.

The same group later found that miR-208, encoded by an intron of the alpha myosin heavy chain (αMHC) gene, is required for cardiomyocyte hypertrophy, fibrosis and expression of αMHC in response to stress and hypothyroidism (van Rooij et al. 2007). The study showed that miR-208 mutant mice failed to undergo stress-induced cardiac remodeling, hypertrophic growth and β MHC upregulation, whereas transgenic expression of miR-208 was sufficient to induce β MHC.

Abdellatif's group reported an array of miRNAs that are differentially and temporally regulated during cardiac hypertrophy (Sayed et al. 2007). They found that miR-1 was singularly downregulated as early as day 1, persisting through day 7, after TAC-induced hypertrophy in a mouse model.

A study from Condorelli's group focused on the role of miR-133 and miR-1 in cardiac hypertrophy with three murine models: TAC mice, transgenic mice with selective cardiac overexpression of a constitutively active mutant of the Akt kinase and human tissues from patients with cardiac hypertrophy (Carè et al. 2007). They showed that cardiac hypertrophy in all three models results in reduced expression levels of both miR-133 and miR-1 in the left ventricle. In vitro overexpression of miR-133 or miR-1 inhibits cardiac hypertrophy. In contrast, suppression of miR-133 induces hypertrophy, which is more pronounced than after stimulation with conventional inducers of hypertrophy. In vivo inhibition of miR-133 by a single infusion of an anti-miRNA antisense oligonucleotide (AMO) against miR-133 causes marked and sustained cardiac hypertrophy.

Cheng et al. (2007) identified 19 deregulated miRNAs in hypertrophic mouse hearts after aortic banding. Knockdown of miR-21 expression via AMO-mediated depletion has a significant negative effect on cardiomyocyte hypertrophy induced by TAC in mice or by angiotensin II or phenylephrine in cultured neonatal cardiomyocytes. Consistently, another independent group identified 17 miRNAs upregulated and 3 miRNAs downregulated in TAC mice and 7 upregulated and 4 downregulated in phenylephrine-induced hypertrophy of neonatal cardiomyocytes. They further showed that inhibition of endogenous miR-21 or miR-18b, which are robustly upregulated, augments hypertrophic growth, while introduction of either of these two miRNAs into cardiomyocytes represses cardiomyocyte hypertrophy (Tatsuguchi et al. 2007).

A study directed to the human heart identified 67 significantly upregulated miRNAs and 43 significantly downregulated miRNAs in failing left ventricles versus normal hearts (Thum et al. 2007). Interestingly, 86.6% of induced miRNAs and 83.7% of repressed miRNAs are regulated in the same direction in fetal and failing heart tissue compared with healthy hearts, consistent with the activation of 'fetal' cardiac genes in heart failure.

Collectively, with respect to hypertrophy, it is evident that in addition to the muscle-specific miRNAs miR-1, miR-133 and miR-208, other miRNAs, including miR-195, miR-21, miR-18b, also play an important role. It appears that multiple miRNAs are involved in cardiac hypertrophy and each of them can independently determine the pathological process.

1.3.4.2 miRNAs and Cardiac Arrhythmias

Arrhythmias are electrical disturbances that can result in irregular heart beating with consequent insufficient pumping of blood. Arrhythmias are often lethal, constituting a major cause for cardiac death in myocardial infarction and heart failure and being one of the most difficult clinical problems. Arrhythmias can occur when there is abnormality in the electrical activities: cardiac conduction, repolarization or automaticity. The electrical activities of the heart are determined by ion channels, the transmembrane proteins embedded across the cytoplasmic membrane of cardiomyocytes. Sodium channels and connexin43 (Cx43) are responsible for excitation generation and inter-cell conduction of excitations, respectively. Calcium channels account for excitation-contraction coupling and contribute to pacemaker activities. Potassium channels govern the membrane potential and rate of membrane repolarization. Pacemaker channels, which carry the nonselective cation current, are critical in generating sinus rhythm under normal conditions and ectopic heart beats as well as in diseased states. Intricate interplays of these ion channels maintain the normal heart rhythm. Dysfunction of any of the ion channels can break the balance rendering arrhythmias.

It has been commonly accepted that regional dispersion of ventricular repolarization is a marker of arrhythmogenicity risk. The spatial heterogeneity of cardiac repolarization is largely due to diversity and varying densities of repolarizing K^+ currents (Liu and Antzelevitch 1995; Verduyn et al. 1997; Szentadrassy et al. 2005). Slowly activating delayed rectifier K^+ current (I_{Ks}) along with its underlying channel proteins KCNQ1 (pore-forming α-subunit) and KCNE1 (auxiliary β-subunit) demonstrates important spatial heterogeneity of distribution, contributing importantly to arrhythmogenicity (Jost et al. 2005). We have experimentally established KCNQ1 and KCNE1 as target genes for repression by miR-133 and miR-1, respectively (Luo et al. 2007). More importantly, we found that the distribution of miR-133 and miR-1 transcripts within the heart is also spatially heterogeneous with the patterns corresponding to the spatial distribution of KCNQ1 and KCNE1 proteins and I_{Ks}.

Abnormal QT interval prolongation is a prominent electrical disorder and has been proposed a predictor of mortality in patients with diabetes mellitus, presumably because it is associated with an increased risk of sudden cardiac death consequent to lethal ventricular arrhythmias. We have previously found that the *ether-a-go-go* related gene (ERG), a long QT syndrome gene encoding a key K^+ channel (I_{Kr}) in cardiac cells, is severely depressed in its expression at the protein level but not at the mRNA level in diabetic subjects (Zhang et al. 2006). To understand the mechanisms underlying the disparate alterations of ERG protein and mRNA, we performed a study on expression regulation of ERG by miRNAs in a rabbit model of diabetes (Xiao et al. 2007). We found remarkable overexpression of miR-133 in diabetic hearts and in parallel, the expression of serum response factor (SRF), which is known to be a transactivator of miR-133, is also found robustly increased. Delivery of exogenous miR-133 into the rabbit myocytes and cell lines produced posttranscriptional repression of ERG, downregulating ERG protein level without altering its transcript level. Correspondingly, forced expression of miR-133 causes substantial depression of I_{Kr}, an effect abrogated by the miR-133 antisense inhibitor. Functional inhibition or gene silencing of SRF downregulated miR-133 expression and increased I_{Kr} density. Repression of ERG by miR-133 likely underlies the differential changes of ERG protein and transcript thereby depression of I_{Kr} and contributes to repolarization slowing thereby QT prolongation and the associated arrhythmias, in diabetic hearts.

One of the most deleterious alterations during myocardial infarction is the occurrence of ischemic arrhythmias (Ghuran and Camm 2001). We found that *miR-1* is overexpressed ($\sim$2.8 fold increase) in the myocardium of individuals with coronary artery disease (CAD) relative to healthy hearts. To explore the mechanisms, we used a rat model of myocardial infarction induced by occlusion of the left anterior descending coronary artery for 12 h that corresponds to the peri-infarction period during which phase II ischemic arrhythmias often occur, which represents a major challenge to our understanding and management of the disorder (Clements-Jewery et al. 2005). We found a similar increase ($\sim$2.6-fold) in miR-1 expression in the ischemic hearts of rats, which is accompanied by exacerbated arrhythmogenesis (Yang et al. 2007). Elimination of miR-1 by an antisense inhibitor in infarcted rat hearts relieved arrhythmogenesis. miR-1 overexpression slows cardiac conduction and depolarizes the cytoplasmic membrane, which is likely the cellular mechanism for the arrhythmogenic potential of miR-1. We further established GJA1, which encodes connexin 43 and KCNJ2, which encodes the K^+ channel subunit Kir2.1 (Wang et al. 1998) as target genes for miR-1. Cx43 is critical for inter-cell conductance and Kir2.1 for setting and maintaining membrane potential. Repression of these proteins by miR-1 explains miR-1-induced slowing of cardiac conduction. We therefore proposed that myocardial infarction upregulates miR-1 expression via some unknown factors, which induces posttranscriptional repression of GJA1 and KCNJ2, resulting in conduction slowing leading to ischemic arrhythmias.

Zhao et al. (2007) determined in vivo miR-1-2 targets, including the cardiac transcription factor, Irx5, which represses KCND2, a potassium channel subunit (Kv4.2) responsible for transient outward K^+ current (I_{to}) (Wang et al. 1999). Their study suggested that the combined loss of Irx5 and Irx4 disrupts ventricular repolarization with a predisposition to arrhythmias. The increase in Irx5 and Irx4 protein levels in miR-1-2 mutants corresponds well with a decrease in KCND2 expression. Clearly, loss-of-function of miR-1 and Dicer mutant embryos affect conductivity through K^+ channels, which supports a central role for miR-1 for fine tuning the regulation of cardiac electrophysiology in pathological and normal conditions.

The pacemaker current I_f, carried by the hyperpolarization-activated channels encoded mainly by the HCN2 and HCN4 genes in the heart, plays an important role in rhythmogenesis. Their expressions reportedly increase in hypertrophic and failing hearts, contributing to arrhythmogenicity under these conditions (Stilli et al. 2001; Fernandez-Velasco et al. 2003). We performed a study on posttranscriptional regulation of HCN2 and HCN4 by miRNAs, experimentally establishing HCN2 mRNA as a target for repression by both miR-1 and miR-133 and HCN4 as a target for miR-1 only (Luo et al. 2008). We further unraveled robust increases in HCN2/HCN4 transcripts and protein levels in a rat model of left ventricular hypertrophy induced by aortic stenosis (narrowing of the abdominal aorta above the left renal artery) and in angiotensin II-induced neonatal cardiomyocyte hypertrophy (Luo et al. 2008). The upregulation of HCN2/HCN4 is accompanied by reduction of miR-1/miR-133. Overexpression of miR-1/miR-133 by transfection prevents largely the overexpression of HCN2/HCN4 in hypertrophic cardiomyocytes. Our data indicate that miR-1/miR-133 act to limit overexpression of HCN2/HCN4 at protein level and downregulation of miR-1/miR-133 underlies partially the abnormal enhancement of HCN2/HCN4 expression in hypertrophic hearts.

1.3.4.3 miRNAs and Vascular Angiogenesis

Proliferative vascular diseases share similar cellular events and molecular mechanisms with cancer and neointimal lesion formation is the pathological basis of proliferative vascular diseases. Neointimal growth is the balance between proliferation and apoptosis of vascular smooth muscle cells (VSMCs). The increased VSMC proliferation or the relative decreased VSMC apoptosis are responsible for neointimal lesion formation. Ji et al. (2007) reported that multiple miRNAs are aberrantly expressed in the vascular wall after angioplasty with the time course changes matching the complex process of neointimal lesion formation, in which multiple genes are accordingly deregulated, as assessed with miRNA microarray. Seven days after balloon injury, 113 of the 140 artery miRNAs are differentially expressed (60 miRNAs are upregulated and 53 miRNAs are downregulated). At 14 days after injury, 110 of the 140 artery miRNAs are differentially expressed (63 up and 47 down), whereas 102 of the 140 artery miRNAs are differentially expressed (55 up and 47 down) at 28 days after angioplasty. Their results indicate that multiple miRNAs are involved in neointimal lesion formation but miR-21 may be of particular significance. miR-21 was found to be one of the most upregulated miRNAs in the vascular wall of balloon-injured rat carotid arteries. Inhibition of miR-21 expression via AMO-mediated miRNA depletion significantly decreases neointima formation after angioplasty by decreasing cell proliferation and increasing cell apoptosis. The authors believe that miRNAs may be a new therapeutic target for proliferative vascular diseases such as atherosclerosis, postangioplasty restenosis, transplantation arteriopathy and stroke.

Knockdown of Dicer by RNAi techniques in human endothelial cells (ECs) alters the expression of several key regulators of endothelial biology and angiogenesis, including TEK/Tie-2, KDR/VEGFR2, Tie-1, endothelial nitric oxide synthase (eNOS) and IL-8 (Suárez et al. 2007). The RNAi-mediated knockdown of Dicer in ECs, results in a significant reduction but not complete loss of mature miRNAs. The net phenotype of Dicer knockdown in ECs is an increase in eNOS protein levels and NO release and decrease in EC growth. Forced expression of miR-222 and miR-221, which are among the highest expressed in ECs, regulates eNOS protein levels after Dicer silencing.

Genetic silencing of Dicer and Drosha significantly reduces capillary sprouting of ECs and tube forming activity (Kuehbacher et al. 2007). Migration of ECs is significantly decreased in Dicer siRNA–transfected cells, whereas Drosha siRNA has no effect. Silencing of Dicer but not of Drosha reduces angiogenesis in vivo. The members of the let-7 family, miR-21, miR-126, miR-221 and miR-222 are highly expressed in endothelial cells. Dicer and Drosha siRNA reduces lef-7f and miR-27b expression. Inhibitors against let-7f and miR-27b also reduce sprout formation indicating that these miRNAs promote angiogenesis by targeting antiangiogenic genes. In silico analysis of predicted targets for let-7 cluster identified the endogenous angiogenesis inhibitor thrombospondin-1. Experimental data confirmed that Dicer and Drosha siRNA significantly increased the expression of thrombospondin-1 (Kuehbacher et al. 2007). These findings indicate that inhibition of Dicer impairs angiogenesis in vitro and in vivo, whereas inhibition of Drosha induces only a minor antiangiogenic effect.

1.3.5 miRNAs and Neuronal Disease

1.3.5.1 miRNAs and Neuronal Development

Efforts to clone tissue-specific miRNAs identified large subsets that are enriched in the brain (Lagos-Quintana et al. 2002; Kim et al. 2004; Kosik and Krichevsky 2005). Expression of many of these miRNAs are regulated during brain development (Miska et al. 2004; Krichevsky et al. 2003) or during development of neurons and astrocytes in culture (Smirnova 2005; Sempere et al. 2004). In situ hybridization analysis with archived human brain samples has revealed the spatial expression patterns of several miRNAs in the brain (Nelson et al. 2006). miR-124a is entirely restricted in its expression to the central nerve system (CNS). miR-315 is transcribed in the brain and a subdomain of the ventral spinal cord; miR-92a and miR-7 are found in cellular subsets of the ventral spinal cord and brain. Other miRNAs like miR-9 and miR-125 are expressed ubiquitously throughout the brain. Another miRNA cluster, miR-13b1/13b2/2c, is expressed throughout the embryonic CNS. And miR-279, the intronic miR-12/283/304 cluster and miR-263b are expressed in the embryonic peripheral nerve system.

Two miRNAs (miR-124 and miR-128) were found to be active primarily in neurons, whereas others (miR-23, miR-26 and miR-29) are preferentially active in astrocytes. miR-124a and miR-9 have been implicated in the decision of a mouse neural precursor to adopt a neuronal or glial fate (Krichevsky et al. 2003), emerging as important regulators of brain morphogenesis. miR-134, a miRNA expressed in hippocampal neurons, has been directly implicated in dendritic spine development (Schratt et al. 2006).

1.3.5.2 miRNAs and Fragile X Syndrome

The neurological disease most frequently associated with miRNAs is human disease fragile X syndrome, which affects 1 in 4,000 males and 1 in 6,000 females of all races and ethnic groups. Fragile X syndrome comprises a family of genetic conditions that commonly manifest with a range of cognitive or intellectual disabilities that vary in severity and which are all caused by mutations in the *FMR1* gene located on the long arm of the X chromosome, which results in the loss of an RNA-binding protein called FMRP (for fragile X mental retardation protein). FMRP seems to influence synaptic plasticity through its role in mRNA transport and translational regulation at synapses. Human FMRP was shown to associate with components of the RNAi machinery (Caudy et al. 2002; Jin et al. 2004; Singh 2007) and genetic interaction of the *Drosophila* homolog dFmr1 with RISC components such as Ago1 has been demonstrated (Jin et al. 2004), suggesting a mechanistic link between miRNA and FMRP translational repression.

1.3.5.3 miRNAs and Tourette's Syndrome

Tourette's syndrome (TS) is a genetically influenced developmental neuropsychiatric disorder characterized by chronic vocal and motor tics. A mutation in the 3′UTR of the SLITRK1 (Slit and Trk-like 1) gene, which is associated with Tourette's syndrome, has been found to enhance repression of the SLITRK mRNA by miR-189, presumably preventing SLITRK1′s promotion of dendritic growth (Abelson et al. 2005). SLITRK1 mRNA and miR-189 have an overlapping expression pattern in brain regions previously implicated in TS. A frame-shift mutation and two independent occurrences of the identical variant in the binding site for miR-189 on chromosome 13q31.1 containing SLITRK1 gene were identified among 174 unrelated probands (Abelson et al. 2005).

1.3.6 miRNAs and Viral Disease

Viruses are obligate intracellular parasites dependent upon the host cellular machinery for their survival and propagation and thus are susceptible to the host gene regulation program. Cellular miRNAs constitute a new layer of endogenous regulatory network to the invading viruses. On the other hand, invading viruses can generate their own miRNAs to target the host genes for their own advantages.

miRNAs are attractive candidates as virally encoded regulators of viral and host cell gene expression due to their small size, their lack of immunogenicity and their remarkable functional flexibility. Known viral miRNAs exclusively derive from dsDNA viruses, mainly of the herpesvirus family, which include a number of human oncogenic viruses like Herpes Simplex virus (HSV), Kaposi Sarcoma Herpes Virus (KSHV) and Epstein Barr virus (EBV) but also from simian polyomaviruses and human adenovirus. These miRNAs have been identified by cDNA cloning of small RNAs from virally infected cells or based on computational prediction and subsequent validation of their expression (Gottwein and Cullen 2008; Scaria 2007; Nair and Zavolan 2006; Sullivan et al. 2005). The number of virus-encoded miRNAs is expected to increase with better computational methods for prediction of putative miRNA precursor candidates and high throughput experimental validation of the candidates.

Some viral miRNAs are expressed from polycistronic transcripts, thereby ensuring their co-regulation. KSHV expresses at least 12 distinct miRNAs during latency from polycistronic transcripts that also encode the KSHV latent proteins (Cai and Cullen 2006). EBV miRNAs are grouped into two clusters, which are differentially expressed in different forms of EBV latency (Cai et al. 2006; Grundhoff et al. 2006; Pfeffer et al. 2004). The BART miRNA cluster contains 20 miRNA precursors within the introns of the BamA rightward transcripts (BARTs). The BHRF1 miRNA cluster, found adjacent to BHRF1 (BamH1 fragment H rightward open reading frame 1), contains three miRNA precursors (Cai et al. 2006; Grundhoff et al. 2006; Pfeffer et al. 2004). Unlike the clustered miRNAs seen in KSHV and EBV, the hCMV and mCMV miRNAs are scattered throughout the viral genome and are expressed in lytically infected fibroblasts (Buck et al. 2007; Dölken et al. 2007; Dunn et al. 2005; Grey et al. 2005; Pfeffer et al. 2005). In most cases, expression of these miRNAs occurs during the early phase of the viral life cycle.

miRNAs of both viral and cellular origins can positively or negatively impact viral replication. Viral miRNAs may directly regulate viral and/or host cell gene expression by engaging in novel regulatory relationships or by mimicking cellular miRNAs and thereby hijacking predefined cellular regulatory networks, including components of the immune system, to benefit the virus. Viruses may utilize cellular miRNAs for their replication and in some cases, the expression of cellular miRNAs may be induced or inhibited to reshape the cellular gene expression environment to the benefit of the virus. On the other hand, the expression of certain cellular miRNAs can directly alter the virus life cycle, limiting virus replication through their interaction with viral mRNAs or because of their cellular functions.

The discovery of viral encoded miRNAs, especially from a family of oncogenic viruses, has attracted immense attention towards the possibility of miRNAs as critical modulators of viral oncogenesis. The host-virus crosstalk mediated by miRNAs, mRNAs and proteins, is complex and involves the different cellular regulatory layers.

1.3.6.1 Viral miRNAs and Host Genes

It is likely that virus-encoded miRNAs can target critical genes associated with disease pathogenesis. Currently, at least eight viral species have been experimentally validated to encode for miRNAs and a majority of them belong to the Herpesvirus family (Pfeffer et al. 2005; Grundhoff et al. 2006; Cai et al. 2005; Samols et al. 2005; Cai and Cullen 2006). The number of miRNA sequences encoded varies widely, ranging from 23 in Epstein–Barr virus (EBV) to two in herpes simplex virus (HSV). The only two functionally validated miRNAs from HSV arise from the latency-associated transcript (LAT) (Cui et al. 2006a; Gupta et al. 2006) and one of them is thought to target transcripts associated with apoptosis in the host (Gupta et al. 2006). Another virus-encoded miRNA effectively modulating cellular defense mechanisms to its benefit has been reported in simian virus 40 (SV40), which is produced during the late phase in the life cycle and helps the virus to evade cytotoxic T cells by targeting early transcripts including those encoding T cell antigens (Sullivan et al. 2005).

Viral miRNAs are not widely conserved and, with the notable exception of viral miRNAs with cellular orthologs, there is no obvious reason to presume that they target conserved regions in cellular mRNAs.

Most viral miRNAs do not share seed homology with cellular miRNAs and therefore likely target novel sites on cellular mRNAs; i.e., they engage in novel, virus-specific regulatory relationships. However, at least some viral miRNAs seem to have evolved to hijack existing and, presumably, highly evolved gene regulatory networks. It was recently reported that KSHV miR-K12-11 is an ortholog of cellular miR-155. miR-K12-11 and miR-155 are identical along their 50-terminal 8 nt, i.e., including the entire seed region (Gottwein et al. 2007; Skalsky et al. 2007).

miR-155, the product of the Bic gene (Eis et al. 2005), is transiently expressed in macrophages, T and B lymphocytes upon treatment with inflammatory stimuli and after T or B cell receptor ligation. Loss of miR-155 expression in knockout mice revealed defects in adaptive immune responses (Haasch et al. 2002; O'Connell et al. 2007; Rodriguez et al. 2007; Thai et al. 2007). Importantly, constitutive expression of miR-155 in B cells is associated with the development of B cell lymphomas in humans, mice and chickens, leading to the classification of miR-155 as an oncomiR. Bic was first identified as a gene that is commonly activated by proviral insertion in avian leukosis virus-induced B cell lymphomas in chicken (Tam et al. 1997). Transgenic expression of miR-155 in the B cell compartment in mice results in the development of B cell lymphomas and miR-155 expression is elevated in several types of human B cell lymphomas, such as Hodgkin's lymphoma, as well as in other cancers (Costinean et al. 2006; van den Berg et al. 2003). While the mechanism by which the constitutive expression of miR-155 in B cells causes tumorigenesis is not understood, the finding that KSHV miR-K12-11 is an ortholog of miR-155 raises the possibility that miR-K12-11 may contribute to the development of KSHV-associated lymphomas. Interestingly, the oncogenic a-herpesvirus MDV-1 encodes a miRNA, miR-M4 that also shares the same 5′-terminal 8 nt, suggesting that this viral miRNA may also function as an ortholog of miR-155 in infected chickens. Remarkably, MDV-2, which lacks a miR-155 ortholog, is not oncogenic in chickens (Schat and Calnek 1978). While MDV-1 also encodes other proteins, such as the meq oncogene that are not present in MDV-2, it is tempting to speculate that MDV-1 miR-M4 may contribute to tumorigenesis in this system. Validated targets for miR-155 include proteins with known roles in B cell function, such as PU.1 but also a protein with known proapoptotic functions, tumor protein p53 inducible nuclear protein 1 (TP53INP1), which has been suggested to contribute to the development of miR-155-expressing pancreatic tumors (Gironella et al. 2007; Vigorito et al. 2007). It is interesting that both KSHV miR-K12-11 and MDV-1 miR-M4 only share the first 8 nts with cellular miR-155. Therefore, it remains possible that functional differences between these miRNAs exist.

A subset of other viral miRNAs also shares sequence homology with cellular miRNAs. MHV68 miR-M1-4 shares the 50-terminal 9 nts with murine miR-151, a miRNA of unknown function. Other seed homologies are more limited and extend only over nts 2–7, corresponding to the minimal miRNA seed region. However, since 6 nt seed base-pairing can be sufficient for the regulation of some targets, these miRNAs may also share targets with their cellular counterparts. The most interesting of these miRNAs are EBV miR-BART5, rLCV miR-rL1-8 and MHV68 miR-M1-7-5p, all of which share perfect seed homology with cellular miR-18a and miR-18b, encoded in the miR-17-92 cluster, which also has oncomiR function (He et al. 2005b).

Another example of viruses subverting host miRNA pathway for their benefit is that hepatitis C virus (HCV) uses the host liver-specific miR-122 that can target the 5′UTR of the viral genome and positively regulates HCV replication (Jopling et al. 2005).

Besides direct interactions between miRNA and viral RNA, viruses can also interfere with the cellular miRNA pathway by competing for the components critical to miRNA biogenesis (Lu and Cullen 2004; Grimm et al. 2006; Andersson et al. 2005). HIV-1 relieves RNAi by diverting TAR RNA-binding protein, a Dicer co-factor, from Dicer to shut down miRNA maturation (Gatignol et al. 2005). More strikingly, viruses can counteract the host miRNA silencing by expressing suppressors to block cellular miRNA pathway. The representatives of such suppressors are PFV-1 Tas protein (Lecellier et al. 2005) and HIV-1 Tat protein (Bennasser et al. 2005).

1.3.6.2 Viral miRNAs and Viral Genes

Viral miRNAs may regulate viral transcripts and thereby determine or fine-tune patterns of viral protein expression. However, experimental evidence has been sparse. Regulation of viral transcripts by viral miRNAs has been seen in HIV where a *nef*-encoded miRNA has been shown to target its own transcript in in vivo and in vitro experiments (Omoto et al. 2004; Omoto and Fujii, 2005). The role of *nef* in disease progression and the inability of *nef* mutants to establish disease leading to long-term nonprogression of disease following HIV infection (Kirchhoff et al. 1995; Salvi et al. 1998) suggests that anti-*nef* miRNAs may have a role in delaying disease progression.

EBV miR-BART2 is fully complementary to the 3′UTR of the mRNA for the EBV DNA polymerase BALF5, which is transcribed antisense to miR-BART2 (Pfeffer et al. 2004). This confers the ability of miR-BART2 to cleavage BALF5 mRNA and explains the detection of an unusual BALF5 transcript upon induction of productive EBV replication (Furnari et al. 1993). Further, BART2-programmed RISC can also cleave BALF5 mRNA (Barth et al. 2008). Since BALF5 is a lytic gene product, it is unclear whether latently expressed miR-BART2 specifically evolved to target BALF5 and whether this interaction is relevant for the maintenance of EBV latency.

A miRNA stem-loop precursor in the genome of the polyomavirus SV40 gives rise to two miRNAs (miR-S1-5p and miR-S1-3p), which are expressed late during viral infection (Sullivan et al. 2005). Both SV40 derived miRNAs are perfectly complementary to early mRNAs transcribed antisense to the pre-miRNA precursor and direct the cleavage of these early transcripts, coding for the large and small T antigens, by RISC. SV40 lacking miRNA expression, due to the selective disruption of the pre-miRNA structure, expressed higher levels of early mRNAs and, consequently, higher levels of large and small T antigen (Sullivan et al. 2005). While the loss of miRNA-mediated downregulation of T antigen was insignificant for virus replication, it increased cellular visibility to SV40-specific cytotoxic T lymphocytes, resulting in a higher susceptibility of SV40 miRNA mutant infected cells to cytotoxic T lymphocytes (Sullivan et al. 2005). The finding that the polyomavirus simian agent 12 (SA12) expresses a homologous miRNA (Cantalupo et al. 2005) suggests that polyomavirus miRNAs may provide a significant advantage to these viruses in vivo (Sullivan et al. 2005).

1.3.6.3 Host miRNAs and Viral Genes

Analysis of miRNAs differentially expressed in HIV infected cells revealed that the miR-17/92 cluster of miRNAs is downregulated following HIV infection. The members of the cluster miR-17-5p and miR-20a repress the translation of PCAF protein that has been previously shown to be a co-factor of Tat in modulating HIV expression. A set of 11 miRNAs are upregulated during HIV infection but the functional role of these miRNAs in modulating viral or host expression is not yet known

(Kumar 2007). It is possible that the host genes modulated by these miRNAs may be important in latency and other pathophysiological changes associated with HIV infection. The cellular miRNA composition in infected cells is likely to indirectly affect viruses, because many pathways that promote or limit viral replication or the survival of infected cells are likely to be regulated by cellular miRNAs.

miR-122 regulates fatty acid and cholesterol biosynthesis, pathways central to liver function and the HCV life cycle (Esau et al. 2006; Krützfeldt et al. 2005). Inhibition of miR-122 function reduced HCV-1 RNA replication, without notable effects on HCV protein translation or RNA stability (Jopling et al. 2005). The importance of miR-122 for the complete HCV life cycle was subsequently confirmed in an HCV-2 cell-culture system (Randall et al. 2007). Furthermore, conservation of the miR-122 binding site in all six HCV genotypes, which otherwise differ by $\sim$30% at the nucleotide level, suggests that the interaction of HCV with miR-122 is relevant in vivo (Jopling et al. 2005). HIV-1 replication in T lymphocytes is enhanced in the context of reduced Dicer and Drosha expression, suggesting that cellular miRNAs may limit HIV-1 replication (Triboulet et al. 2007). They further showed that miRNA expression from the cellular miR-17-92 cluster, comprising miR-17-5p, miR-18, miR-19a, miR-20a, miR-19b-1 and miR-92-1, is downregulated in HIV-1-infected T lymphocytes by an unknown mechanism.

On the other hand, replication of wild-type viruses can also be inhibited by endogenous cellular miRNAs, raising the question of whether specific cellular miRNAs may have specifically evolved to limit virus replication. Lecellier et al. (2005) showed that endogenous cellular miR-32, expressed in HeLa and BHK21 cells, can limit the replication of the retrovirus primate foamy virus (PFV) in cell culture through an interaction with a poorly conserved region in the 3′ portion of the PFV genome (Lecellier et al. (2005). Dicer knockout mice are hypersensitive to infection by vesicular stomatitis virus (VSV), a negative-strand RNA virus. This effect is at least partially due to the loss of miR-24 and miR-93 expression in these mice. Both miRNAs downregulate viral mRNAs and mutant VSV, in which miR-24 and miR-93 binding is abolished. The transfection of HCV replicon-containing hepatocytes with synthetic miRNAs miR-196, miR-296, miR-351, miR-431 and miR-448 results in a 2- to 5-fold reduction in HCV RNA accumulation. The antiviral effect of the transfected miRNA is correlated with the presence of matches to these miRNAs in the viral genome (Pedersen et al. 2007).

1.3.7 miRNAs and Metabolic Disorders

Metabolic disorders such as diabetes and obesity have become recognized as a major challenge to global health. Several studies with combined bioinformatics and bench research have provided evidence that miRNAs affect pathways that are fundamental for metabolic control involving lipid, amino acid, glucose homeostasis and the closely related process lifespan (Boehm and Slack 2006; Cuellar and McManus 2005; Krützfeldt and Stoffel 2006; Poy et al. 2007).

These studies described below open the possibility that miRNAs may contribute to common metabolic diseases and point to understanding the metabolic and lifespan regulatory roles of these novel gene regulators will undoubtedly further our understanding of the complex genetic networks that control metabolism and will also provide us with novel therapeutic opportunities based on targeting of miRNAs.

1.3.7.1 miRNAs and Lifespan

The first evidence for participation of miRNAs in metabolism came from a forward genetic screen in the fruit fly *Drosophila melanogaster*. Xu et al. (2003) found that loss of miR-14 doubles the amount of total body triacylglycerides. Similarly, miR-278 is prominently expressed in the fat body of flies and homozygous mutations for miR-278 have a smaller fat body and reduces ratio of total body triglycerides to total protein. This phenotype is rescued by miR-278 expression (Teleman et al. 2005). In *C. elegans*, overexpression of lin-4 leads to a lengthened lifespan, whereas animals with loss-of-function mutation in lin-4 displays a lifespan that is significantly shorter than that of the wild-type (Boehm and Slack 2005).

Though these results cannot directly be related to mammalian and human tissues. However, it is still possible that miR-14 and miR-278 regulate the expression of evolutionarily conserved target genes, which may shed light on novel pathways affecting insulin signaling and energy homeostasis.

1.3.7.2 miRNAs and Glucose Homeostasis

The best-known study on miRNA regulation of metabolism in mammals may be from the report showing the role of a miRNA in insulin secretion and glucose homeostasis. Poy et al. (2007) demonstrated that mouse insulin secretion is regulated by the pancreatic miRNA miR-375, an evolutionarily conserved islet-specific miRNA. Overexpression of miR-375 decreases insulin secretion, while inhibiting miR-375 function increases insulin release from pancreatic β-cells. Myotrophin is evidently the target of miR-375 and this miRNA:mRNA interaction regulates the secretion of insulin. Also striking is that in addition to miR-375, the authors found another 67 miRNAs expressed in β-cells and future studies should determine if these other miRNAs are involved in pancreatic β-cell development or in the regulation of insulin production or secretion. The regulation of exocytosis machinery is probably a widespread phenomenon because many miRNAs are predicted to target exocytosis-related proteins (Abderrahmani et al. 2006). Plaisance et al. (2006) showed that miR-9 causes a reduction in exocytosis in pancreas (insulin secreting beta cells) that is elicited by stimuli including glutamate.

1.3.7.3 miRNAs and Glucose Homeostasis

miR-122, the predominant miRNA in the liver, has been shown to regulate cholesterol and lipid homeostasis in two independent studies. This miRNA is abundantly expressed in human and rodent liver tissue, with estimates ranging from 50,000 to 80,000 copies per cell. Using a modified anti-miRNA inhibitor, termed antagomiR, Krützfeldt et al. 2005) studied the effect of miR-122 on glucose and lipid metabolism in mice. The antagomiR against miR-122 induces efficient degradation of miR-122 in the liver and leads to a significant decrease in plasma cholesterol levels. Esau et al. 2006) extended these findings using a differently modified anti-miRNA inhibitor and a longer treatment protocol. They found that the miR-122 inhibitor-treated mice exhibited decreased cholesterol and triglyceride levels without changes in plasma glucose concentrations. Isolated hepatocytes from treated mice have decreased hepatic fatty-acid synthesis and sterol synthesis as well as increased fatty-acid oxidation.

There is also emerging evidence that miRNAs might play a role in differentiation of insulin-sensitive organs such as the adipocyte and muscle (Esau et al. 2004). Knockdown of miR-9 and miR-143 by their respective AMOs inhibits the differentiation process as assessed by a reduction in triglyceride accumulation and decreases expression of adipocyte-specific genes. Furthermore, miR-143 expression levels are higher in adipocytes than in preadipocytes.

1.3.7.4 miRNAs and Lipid Metabolism

In addition, evidence indicates that miRNAs may also be involved in regulating pathways in amino acid metabolism. Mersey et al. (2005) showed that the miR-29b controls the amount of the branched-chain α-ketoacid dehydrogenase (BCKD) complex in mammalian cells (HEK293), which catalyzes the first irreversible step in branched-chain amino acid catabolism. The branched-chain amino acids including leucine, isoleucine and valine are amino acid components in nearly all proteins and leucine can stimulate protein synthesis and act as a stimulus for insulin secretion.

1.3.8 miRNA and Epigenetics

Epigenetics is defined as mitotically and meiotically heritable changes in gene expression that do not involve a change in the DNA sequence (Chuang and Jones 2007; Saetrom 2007). Two major areas of epigenetics – DNA methylation and histone modifications – are known to have profound effects on controlling gene expression. DNA methylation is involved in maintaining a normal expression pattern of mammalian cells. DNA methylation occurs almost exclusively on a cytosine in a CpG dinucleotide and is achieved by the addition of a methyl group to the position 5 of a cytosine ring mediated by DNMTs. Histone modifications,especially the posttranslational modifications of amino-terminal tail domains, are also important epigenetic mechanisms in controlling gene expression. Certain histone modifications, such as histone acetylation, are associated with active gene transcription, whereas others such as the methylation of histone H3 lysine 9 is an indicator of condensed and inactive chromatin. DNA methylation and histone modification play critical roles in chromatin remodeling and general regulation of gene expression in mammalian development and aberrant hypermethylation can result in human diseases, such as cancer.

1.3.8.1 miRNAs and DNA Methylation

miRNAs can be involved in establishing DNA methylation. miR-165 and miR-166 have been shown to be required for the methylation at the *PHABULOSA (PHB)* gene in *Arabidopsis*. They interact with the newly processed *PHB* mRNA to change the chromatin of the template *PHB* gene (Bao et al. 2004). This presents a novel mechanism by which miRNAs control gene expression in addition to the posttranscriptional repression. Similar findings in mammalian cells are yet to be demonstrated.

1.3.8.2 miRNAs and histone modifications

In addition, miRNAs may regulate chromatin structure by regulating key histone modifiers. miR-140, which is cartilage specific, can target histone deacetylase 4 in mice (Tuddenham et al. 2006). Costa et al. (2006) suggested that miRNAs may be involved in meiotic silencing of unsynapsed chromatin in mice. Taken together, miRNAs can be considered important players in the epigenetic control of gene expression.

1.3.8.3 DNA Methylation/Histone Modifications and miRNAs

On the other hand, DNA methylation and histone modifications can also affect expression of miRNAs. Saito et al. (2006) showed that $\sim$5% of human miRNAs are upregulated more than threefold by treatment of T24 bladder cancer cells with DNA demethylating agent 5-Aza-CdR and histone deacetylase inhibitor 4-phenylbutyric acid. In particular, miR-127, which is embedded in a CpG island, is remarkably induced by a decrease in DNA methylation levels and an increase in active histone marks around the promoter region of the miR-127 gene. These findings suggest that some miRNA genes are controlled by epigenetic alterations in their promoter regions and can be activated by inhibitors of DNA methylation and histone deacetylase. miR-127 is highly expressed in normal prostate and bladder tissues but is remarkably downregulated or silenced in the corresponding tumors. In addition, the proto-oncogene *BCL6* has been identified as one of the targets of miR-127, suggesting that miR-127 has a role as a tumor suppressor. Scott et al. (2006) also showed that treatment of breast cancer cell line SKBr3 with histone deacetylase inhibitor LAQ824 led to a rapid change in miRNA expression profile.

1.4 miRNAs as Therapeutic Targets

The main purpose of the above sections is to convey the key message of this chapter: miRNAs are viable therapeutic targets for human diseases. The rationale for using miRNAs as potential therapeutic targets is based on the following facts.

1. Aberrant miRNA expression is a common feature of human multigenic diseases and could be used as a prognostic biomarker for disease diagnosis.
2. Particular miRNAs are involved in, as causal factors or key contributors, particular types of human diseases and could serve as novel therapeutic targets for disease treatment.
3. The level of miRNA expression responds to physiological stimuli and is susceptible to drug intervention.
4. Small size (18–26 nucleotides in length) of miRNAs makes them very attractive for drug development.
5. A miRNA can be viewed as a regulator of a cellular function or a cellular program, not of a single gene (my hypothesis) (Wang et al. 2008). We can conclude that miRNAs represent a class of genes with a great potential for use in diagnosis, prognosis and therapy and are a new frontier for molecular medicine. Indeed, miRNAs as a valid diagnostic tool and therapeutic targets for modern molecular therapy of human disease have yielded promising data and encourage boosting interests from pharmaceuticals (Yang et al. 2007; Wurdinger and Costa 2007; Soifer et al. 2007; Zhang 2008; Esau and Monia 2007; Mack 2007).

The wide use of RNAi (RNA interference) methodologies employing siRNAs (small interference RNAs) or shRNAs (short RNAs) to silence single target genes has contributed to recent advances in molecular biology and molecular therapy (Moffat and Sabatini 2006; Behlke 2006). Currently, siRNAs are being tested in experimental therapy for a variety of diseases and a few are starting to enter clinical trials (Vidal et al. 2005). The use of siRNAs with perfect pairing to the target mRNA bears several drawbacks. First, off-target effects or the RNAi-mediated silencing of genes other than the target gene have been frequently observed (Jackson and Linsley 2004; Jackson and Linsley 2003). These can be due to the miRNA-like action of siRNAs through imperfect binding of the particular siRNA to nontarget mRNAs (Jackson et al. 2006). That is, siRNAs can act by the mechanism of miRNA actions when they are partially complementary to nontarget mRNAs. Second, elicitation of the interferon response by high levels of siRNAs and the clogging of export of mRNAs from the nucleus by overloading with shRNAs have recently been described (Birmingham et al. 2006; Fedorov et al. 2006; Lin et al. 2005; Grimm et al. 2006). Moreover, as various microarray screens indicate, in unhealthy tissues there can be an imbalanced gene expression pattern involving many genes and many of the human diseases are multigenic (Chung et al. 2002; Hoheisel 2006), targeting a single gene with RNAi may not be effective as therapeutic intervention. Though RNAi has proven its usefulness as a tool with good success in knocking down of particular single genes (Mello and Conte 2004), miRNAs provide an alternative approach towards targeting multigenic disease conditions by the mechanism of multiple target regulation (Wurdinger and Costa 2007).

To this end, Wurdinger and Costa (2007) proposed the 'one hit, multiple targets' concept, based on a key property that one miRNA can downregulate multiple target proteins by interacting with different target mRNAs (Lim et al. 2005), which is essentially a homolog to the original 'one agent, multiple targets' concept that I proposed in 2006 (Gao et al. 2006). This concept reasons that if the primary molecular defect of the disease is in miRNAs or in the miRNA pathway and, as a consequence, the expression of the protein-coding mRNA targets is deregulated, one could intervene by 'normalizing' or 'correcting' the miRNA expression. Here I reinforce this idea: one could also intervene by 'normalizing' or 'correcting' the miRNA function, biogenesis and stability.

1.4.1 Strategies for Therapeutic Modulation of miRNAs

This section aims to lead readers to enter the 'palace' of miRNA interference (miRNAi) by giving a brief introduction to the strategies and approaches that have been experimentally validated as miRNA research tools and therapeutic modulation of miRNAs. The concept of miRNAi pertinent to these strategies and approaches are described in Chap. 2. The detailed descriptions of each individual technology of miRNAi for therapeutic modulation of miRNAs are given in the Chaps. 3–13.

1.4.1.1 Normalizing miRNA Expression

'Normalizing or correcting aberrant expression of miRNAs being implicated in diseased conditions' means upregulating expression of downregulated miRNAs and downregulating expression of upregulated miRNAs. Several methods have proven valid for achieving this goal. First, forced transient expression of exogenously supplied canonical miRNAs by lipid-mediated transfection or virus-mediated infection is the most commonly used, successful way of upregulating miRNAs in cell culture conditions. Second, long-term overexpression of miRNAs by genetic engineering in the transgenic mouse model has been tested in several occasions (Lu et al. 2007; van Rooij et al. 2006, 2007; Lin et al. 2006). Third, knockout of miRNA-coding genes in mice to create null expression of the target miRNA (Zhao et al. 2007). Finally, transcriptional regulation of miRNA expression is definitely an alternative for manipulating miRNA levels. Interest in understanding the genomic structures and the regulatory elements of miRNAs has been rapidly increasing. A few mammalian miRNAs have recently been characterized for their promoter regions (Zhao et al. 2005; Chen et al. 2006; Liu et al. 2007; Taganov et al. 2006; Saito et al. 2006; Hino et al. 2008; Fujita et al. 2008; Yin et al. 2008; Motsch et al. 2007; Corney et al. 2007).

1.4.1.2 Modulation of miRNA Biogenesis

As introduced in the Sect. 1.1.1 of this chapter, miRNA biogenesis requires processing by a series of steps involving several enzymes and co-factors. In theory, altering any of these enzymes and co-factors should be able to induce changes of miRNA biosynthesis thereby changes of miRNA level in a cell. Thus far, the most successfully examined method has been the tissue-restricted Dicer deletion, which results in significant loss of miRNAs and change of related phenotypes (Otsuka et al. 2007, 2008; Murchison et al. 2007; Chen et al. 2008a; Hayashi et al. 2008; Mudhasani et al. 2008). siRNAs targeting Drosha or Dicer have also been tested (Kuehbacher et al. 2007; Gillies and Lorimer 2007). Ago2-deficient mice have been generated to study the function of miRNAs and miRNAs:RISC interactions (Hayashi et al. 2008; Morita et al. 2007; O'Carroll et al. 2007).

1.4.1.3 Modulation of miRNA Stability

Anti-miRNA antisense inhibitor oligonucleotides (AMO or ASO) and chemically modified AMO named antagomiRs are probably the most successful methods of inhibiting miRNA action (Esau 2008; Krützfeldt et al. 2005). Recent studies indicate that these molecules can also affect the stability of miRNAs by inducing degradation of the targeted miRNAs; this mode of effect significantly decreases miRNA level.

1.4.1.4 Modulation of miRNA Function

One can alter the actions of miRNAs without changing miRNA expression and stability. The AMO and antagomiR techniques mentioned above are highly efficient ways of preventing miRNAs from performing their posttranscriptional repressive function and represent a most promising class of molecules for drug development for miRNA-based molecular therapy. We have also developed several new technologies for interfering miRNA functions and they are described in the following chapters.

1.4.2 *Approaches for Therapeutic Modulation of miRNAs*

1.4.2.1 Gain-of-function of miRNAs

Gain-of-function of miRNAs refers to any manipulations on expression, biogenesis, stability, or function of miRNAs, leading to enhancement of function of miRNAs. Gain-of-function technology is an indispensible approach in miRNA research and gene silencing by miRNA mechanisms.

1.4.2.2 Loss-of-function of miRNAs

Loss-of-function of miRNAs refers to any negative intervention resulting in inhibition of expression, biogenesis, stability, or function of miRNAs. Loss-of-function technology has been one of the most popular knockdown/knockout tools for the study of gene function in cell and developmental biology.

References

Abderrahmani A, Plaisance V, Lovis P, Regazzi R (2006) Mechanisms controlling the expression of the components of the exocytotic apparatus under physiological and pathological conditions. Biochem Soc Trans 34:696–700.

Abelson JF, Kwan KY, O'Roak BJ, Baek DY, Stillman AA, Morgan TM, Mathews CA, Pauls DL, Rasin MR, Gunel M, Davis NR, Ercan-Sencicek AG, Guez DH, Spertus JA, Leckman JF, Dure LS 4th, Kurlan R, Singer HS, Gilbert DL, Farhi A, Louvi A, Lifton RP, Sestan N, State MW (2005) Sequence variants in SLITRK1 are associated with Tourette's syndrome. Science 310:317–320.

Alvarez-Garcia I, Miska EA (2005) MicroRNA functions in animal development and human disease. Development 132:4653–4662.

Ambros V, Bartel B, Bartel DP, Burge CB, Carrington JC, Chen X, Dreyfuss G, Eddy SR, Griffiths-Jones S, Marshall M, Matzke M, Ruvkun G, Tuschl T (2003a) A uniform system for microRNAs annotation. RNA 9:277–279.

Ambros V, Lee RC, Lavanway A, Williams PT, Jewell D (2003b) MicroRNAs and other tiny endogenous RNAs in *C. elegans*. Curr Biol 13:807–818.

Ambros V (2004) The functions of animal microRNAs. Nature 431:350–355.

Andersson MG, Haasnoot PC, Xu N, Berenjian S, Berkhout B, Akusjärvi G (2005) Suppression of RNA interference by adenovirus virus-associated RNA. J Virol 79:9556–9565.

Aparicio O, Razquin N, Zaratiegui M, Narvaiza I, Fortes P (2006) Adenovirus virus-associated RNA is processed to functional interfering RNAs involved in virus production. J Virol 80:1376–1384.

Arisawa T, Tahara T, Shibata T, Nagasaka M, Nakamura M, Kamiya Y, Fujita H, Hasegawa S, Takagi T, Wang FY, Hirata I, Nakano H (2007) A polymorphism of microRNA 27a genome region is associated with the development of gastric mucosal atrophy in Japanese male subjects. Dig Dis Sci 52:1691–1697.

Bao N, Lye KW, Barton MK (2004) MicroRNA binding sites in Arabidopsis class III HD-ZIP mRNAs are required for methylation of the template chromosome. Dev Cell 7:653–662.

Barciszewska MZ, Szymaski M, Erdmann VA, Barciszewski J (2000) 5S ribosomal RNA. Biomacromolecules 1:297–302.
Bartel DP (2004) MicroRNAs: Genomics, biogenesis, mechanism, and function. Cell 116: 281–297.
Barth S, Pfuhl T, Mamiani A, Ehses C, Roemer K, Kremmer E, Jäker C, Höck J, Meister G, Grässer FA (2008) Epstein-Barr virus-encoded microRNA miR-BART2 down-regulates the viral DNA polymerase BALF5. Nucleic Acids Res 36:666–675.
Bauersachs J, Thum T (2007) MicroRNAs in the broken heart. Eur J Clin Invest 37:829–833.
Behlke MA (2006) Progress towards in vivo use of siRNAs. Mol Ther 13:644–670.
Bennasser Y, Le SY, Benkirane M, Jeang KT (2005) Evidence that HIV-1 encodes an siRNA and a suppressor of RNA silencing. Immunity 22:607–169.
Berezikov E, Chung WJ, Willis J, Cuppen E, Lai EC (2007) Mammalian mirtron genes. Mol Cell 28:328–336.
Bernstein E, Kim SY, Carmell MA, Murchison EP, Alcorn H, Li MZ, Mills AA, Elledge SJ, Anderson KV, Hannon GJ (2003) Dicer is essential for mouse development. Nat Genet 35:215–217.
Bertino JR, Banerjee D, Mishra PJ (2007) Pharmacogenomics of microRNA: A miRSNP towards individualized therapy. Pharmacogenomics 8:1625–1627.
Birmingham A, Anderson EM, Reynolds A, Ilsley-Tyree D, Leake D, Federov Y et al (2006) 30 UTR seed matches, but not overall identity, are associated with RNAi off-targets. Nat Methods 3:199–204.
Boehm M, Slack FJ (2005) A developmental timing microRNA and its target regulate life span in C. elegans. Science 310:1954–1957.
Boehm M, Slack FJ (2006) MicroRNA control of lifespan and metabolism. Cell Cycle 5:837–840.
Bohnsack MT, Czaplinski K, Gorlich D (2004) Exportin 5 is a RanGTP-dependent dsRNA-binding protein that mediates nuclear export of pre-miRNAs. RNA 10:185–191.
Bommer GT, Gerin I, Feng Y, Kaczorowski AJ, Kuick R, Love RE, Zhai Y, Giordano TJ, Qin ZS, Moore BB, MacDougald OA, Cho KR, Fearon ER (2007) p53-mediated activation of miRNA34 candidate tumor-suppressor genes. Curr Biol 17:1298–1307.
Bottoni A, Piccin D, Tagliati F, Luchin A, Zatelli MC, degli Uberti EC (2005) miR-15a and miR-16-1 down-regulation in pituitary adenomas. J Cell Physiol 204:280–285.
Brantl S (2002) Antisense-RNA regulation and RNA interference. Biochimica et Biophysica Acta 1575:15–25.
Brantl S (2007) Regulatory mechanisms employed by cis-encoded antisense RNAs. Curr Opin Microbiol 10:102–109.
Breaker RR (2008) Complex riboswitches. Science 319:1795–1797.
Brennecke J, Stark A, Russell RB, Cohen SM (2005) Principles of microRNA-target recognition. PLoS Biol 3:e85.
Buck AH, Santoyo-Lopez J, Robertson KA, Kumar DS, Reczko M, Ghazal P (2007) Discrete clusters of virus-encoded micrornas are associated with complementary strands of the genome and the 7.2-kilobase stable intron in murine cytomegalovirus. J Virol 81:13761–13770.
Burnside J, Bernberg E, Anderson A, Lu C, Meyers BC, Green PJ, Jain N, Isaacs G, Morgan RW (2006) Marek's disease virus encodes MicroRNAs that map to meq and the latency-associated transcript. J Virol 80:8778–8786.
Cai X, Cullen BR (2006) Transcriptional origin of Kaposi's sarcoma-associated herpesvirus microRNAs. J Virol 80:2234–2242.
Cai X, Schäfer A, Lu S, Bilello JP, Desrosiers RC, Edwards R, Raab-Traub N, Cullen BR (2006) Epstein-Barr virus microRNAs are evolutionarily conserved and differentially expressed. PLoS Pathog 2:e23.
Calin GA, Croce CM (2006) MicroRNA-Cancer connection: The beginning of a new tale. Cancer Res 66:7390–7394.
Calin GA, Dumitru CD, Shimizu M, Bichi R, Zupo S, Noch E, Aldler H, Rattan S, Keating M, Rai K, Rassenti L, Kipps T, Negrini M, Bullrich F, Croce CM (2002) Frequent deletions

and down-regulation of microRNA genes miR15 and miR16 at 13q14 in chronic lymphocytic leukemia. Proc Natl Acad Sci USA 99:15524–15529.

Calin GA, Ferracin M, Cimmino A, Di Leva G, Shimizu M, Wojcik SE, Iorio MV, Visone R, Sever NI, Fabbri M, Iuliano R, Palumbo T, Pichiorri F, Roldo C, Garzon R, Sevignani C, Rassenti L, Alder H, Volinia S, Liu CG, Kipps TJ, Negrini M, Croce CM (2005) A MicroRNA signature associated with prognosis and progression in chronic lymphocytic leukemia. N Engl J Med 353:1793–1801.

Calin GA, Liu CG, Sevignani C, Ferracin M, Felli N, Dumitru CD, Shimizu M, Cimmino A, Zupo S, Dono M, Dell'Aquila ML, Alder H, Rassenti L, Kipps TJ, Bullrich F, Negrini M, Croce CM (2004a) MicroRNA profiling reveals distinct signatures in B cell chronic lymphocytic leukemias. Proc Natl Acad Sci USA 101:11755–11760.

Calin GA, Sevignani C, Dumitru CD, Hyslop T, Noch E, Yendamuri S, Shimizu M, Rattan S, Bullrich F, Negrini M, Croce CM (2004b) Human microRNA genes are frequently located at fragile sites and genomic regions involved in cancers. Proc Natl Acad Sci USA 101:2999–3004.

Cantalupo P, Doering A, Sullivan CS, Pal A, Peden KW, Lewis AM, Pipas JM (2005) Complete nucleotide sequence of polyomavirus SA12. J Virol 79:13094–13104.

Carè A, Catalucci D, Felicetti F, Bonci D, Addario A, Gallo P, Bang ML, Segnalini P, Gu Y, Dalton ND, Elia L, Latronico MV, Høydal M, Autore C, Russo MA, Dorn GW, Ellingsen O, Ruiz-Lozano P, Peterson KL, Croce CM, Peschle C, Condorelli G (2007) MicroRNA-133 controls cardiac hypertrophy. Nat Med 13:613–618.

Carmell MA, Xuan Z, Zhang MQ, Hannon GJ (2002) The Argonaute family: Tentacles that reach into RNAi, developmental control, stem cellmaintenance, and tumorigenesis. Genes Dev 16:2733–2742.

Caudy AA, Myers M, Hannon GJ, Hammond SM (2002) Fragile X-related protein and VIG associate with the RNA interference machinery, Genes Dev 16:2491–2496.

Chan JA, Krichevsky AM, Kosik KS (2005) MicroRNA-21 is an antiapoptotic factor in human glioblastoma cells. Cancer Res 65:6029–6033.

Chang TC, Wentzel EA, Kent OA, Ramachandran K, Mullendore M, Lee KH, Feldmann G, Yamakuchi M, Ferlito M, Lowenstein CJ, Arking DE, Beer MA, Maitra A, Mendell JT (2007) Transactivation of miR-34a by p53 broadly influences gene expression and promotes apoptosis. Mol Cell 26:745–752.

Chen JF, Mandel EM, Thomson JM, Wu Q, Callis TE, Hammond SM, Conlon FL, Wang DZ (2006) The role of microRNA-1 and microRNA-133 in skeletal muscle proliferation and differentiation. Nat Genet 38:228–233.

Chen JF, Murchison EP, Tang R, Callis TE, Tatsuguchi M, Deng Z, Rojas M, Hammond SM, Schneider MD, Selzman CH, Meissner G, Patterson C, Hannon GJ, Wang DZ (2008a) Targeted deletion of Dicer in the heart leads to dilated cardiomyopathy and heart failure. Proc Natl Acad Sci USA 105:2111–2116.

Chen K, Song F, Calin GA, Wei Q, Hao X, Zhang W (2008b) Polymorphisms in microRNA targets: A gold mine for molecular epidemiology. Carcinogenesis 29:1306–1311.

Cheng AM, Byrom MW, Shelton J, Ford LP (2005) Antisense inhibition of human miRNAs and indications for an involvement of miRNA in cell growth and apoptosis. Nucleic Acids Res 33:1290–1297.

Cheng Y, Ji R, Yue J, Yang J, Liu X, Chen H, Dean DB, Zhang C (2007) MicroRNAs are aberrantly expressed in hypertrophic heart. Do they play a role in cardiac hypertrophy? Am J Pathol 170:1831–1840.

Chendrimada TP, Gregory RI, Kumaraswamy E, Norman J, Cooch N, Nishikura K, Shiekhattar R (2005) TRBP recruits the Dicer complex to Ago2 for microRNA processing and gene silencing. Nature 436:740–744.

Chung CH, Bernard PS, Perou CM (2002) Molecular portraits and the family tree of cancer. Nat Genet 32:533–540.

Chuang JC, Jones PA (2007) Epigenetics and microRNAs. Pediatr Res 61:24R–29R.

Ciafre SA, Galardi S, Mangiola A, Ferracin M, Liu CG, Sabatino G, Negrini M, Maira G, Croce CM, Farace MG (2005) Extensive modulation of a set of microRNAs in primary glioblastoma. Biochem Biophys Res Commun 334:1351–1358.

Cimmino A, Calin GA, Fabbri M, Iorio MV, Ferracin M, Shimizu M, Wojcik SE, Aqeilan RI, Zupo S, Dono M, Rassenti L, Alder H, Volinia S, Liu CG, Kipps TJ, Negrini M, Croce CM (2005) miR-15 and miR-16 induce apoptosis by targeting BCL2. Proc Natl Acad Sci USA 102:13944–13949.

Clements-Jewery H, Hearse DJ, Curtis MJ (2005) Phase 2 ventricular arrhythmias in acute myocardial infarction: A neglected target for therapeutic antiarrhythmic drug development and for safety pharmacology evaluation. Br J Pharmacol 145:551–564.

Clop A, Marcq F, Takeda H, Pirottin D, Tordoir X, Bibé B, Bouix J, Caiment F, Elsen JM, Eychenne F, Larzul C, Laville E, Meish F, Milenkovic D, Tobin J, Charlier C, Georges M. (2006) A mutation creating a potential illegitimate microRNA target site in the myostatin gene affects muscularity in sheep. Nat Genet 38:813–818.

Corney DC, Flesken-Nikitin A, Godwin AK, Wang W, Nikitin AY (2007) MicroRNA-34b and MicroRNA-34c are targets of p53 and cooperate in control of cell proliferation and adhesion-independent growth. Cancer Res 67:8433–8438.

Corsten MF, Miranda R, Kasmieh R, Krichevsky AM, Weissleder R, Shah K (2007) MicroRNA-21 knockdown disrupts glioma growth in vivo and displays synergistic cytotoxicity with neural precursor cell delivered S-TRAIL in human gliomas. Cancer Res 67:8994–9000.

Costa FF (2007) Non-coding RNAs: Lost in translation? Gene 386:1–10.

Costa Y, Speed RM, Gautier P, Semple CA, Maratou K, Turner JM, Cooke HJ (2006) Mouse MAELSTROM: The link between meiotic silencing of unsynapsed chromatin and microRNA pathway? Hum Mol Genet 15:2324–2334.

Costinean S, Zanesi N, Pekarsky Y, Tili E, Volinia S, Heerema N, Croce CM (2006) Pre-B cell proliferation and lymphoblastic leukemia/high-grade lymphoma in E(mu)-miR155 transgenic mice. Proc Natl Acad Sci USA 103:7024–7029.

Cox DN, Chao A, Baker J, Chang L, Qiao D, Lin H (1998) A novel class of evolutionarily conserved genes defined by piwi are essential for stem cell self-renewal. Genes Dev 12:3715–3727.

Cuellar TL, McManus MT (2005) MicroRNAs and endocrine biology. J Endocrinol 187:327–332.

Cui C, Griffiths A, Li G, Silva LM, Kramer MF, Gaasterland T, Wang XJ, Coen DM (2006a) Prediction and identification of herpes simplex virus 1-encoded microRNAs. J Virol 80:5499–5508.

Cui Q, Yu Z, Purisima E, Wang E (2006b) Principles of microRNA regulation of a human cellular signaling network. Mol Systems Biol 2:46–52.

Cullen BR (2004) Transcription and processing of human microRNA precursors. Mol Cell 16:861–865.

Cummins JM, He Y, Leary RJ, Pagliarini R, Diaz LA Jr, Sjoblom T, Barad O, Bentwich Z, Szafranska AE, Labourier E, Raymond CK, Roberts BS, Juhl H, Kinzler KW, Vogelstein B, Velculescu VE (2006) The colorectal microRNAome. Proc Natl Acad Sci USA 103:3687–3692.

Dalmay T (2008) MicroRNAs and cancer. J Inter Med 263:366–375.

Darnell DK, Kaur S, Stanislaw S, Konieczka JH, Yatskievych TA, Antin PB (2006) MicroRNA expression during chick embryo development. Dev Dyn 235:3156–3165.

Davis E, Caiment F, Tordoir X, Cavaillé J, Ferguson-Smith A, Cockett N, Georges M, Charlier C (2005) RNAi-mediated allelic trans-interaction at the imprinted Rtl1/Peg11 locus. Curr Biol 15:743–749.

Denli AM, Tops BB, Plasterk RH, Ketting RF, Hannon GJ (2004) Processing of primary microRNAs by the Microprocessor complex. Nature 432:231–235.

Deshpande G, Calhoun G, Schedl P (2005) Drosophila Argonaute-2 is required early in embryogenesis for the assembly of centric/centromeric heterochromatin, nuclear division, nuclear migration, and germ-cell formation. Genes Dev 19:1680–1685.

Diederichs S, Haber DA (2006) Sequence variations of microRNAs in human cancer: Alterations in predicted secondary structure do not affect processing. Cancer Res 66:6097–6104.

Dölken L, Perot J, Cognat V, Alioua A, John M, Soutschek J, Ruzsics Z, Koszinowski U, Voinnet O, Pfeffer S (2007) Mouse cytomegalovirus microRNAs dominate the cellular small RNA profile during lytic infection and show features of posttranscriptional regulation. J Virol 81:13771–13782.

Duan R, Pak C, Jin P (2007) Single nucleotide polymorphism associated with mature miR-125a alters the processing of pri-miRNA. Hum Mol Genet 16:1124–1131.

Dunn W, Trang P, Zhong Q, Yang E, van Belle C, Liu F (2005) Human cytomegalovirus expresses novel microRNAs during productive viral infection. Cell Microbiol 7:1684–1695.

Eis PS, Tam W, Sun L, Chadburn A, Li Z, Gomez MF, Lund E, Dahlberg JE (2005) Accumulation of miR-155 and BIC RNA in human B cell lymphomas. Proc Natl Acad Sci USA 102:3627–3632.

Esau CC (2008) Inhibition of microRNA with antisense oligonucleotides. Methods 44:55–60.

Esau C, Davis S, Murray SF, Yu XX, Pandey SK, Pear M, Watts L, Booten SL, Graham M, McKay R, Subramaniam A, Propp S, Lollo BA, Freier S, Bennett CF, Bhanot S, Monia BP (2006) miR-122 regulation of lipid metabolism revealed by in vivo antisense targeting. Cell Metab 3:87–98.

Esau C, Kang X, Peralta E, Hanson E, Marcusson EG, Ravichandran LV, Sun Y, Koo S, Perera RJ, Jain R, Dean NM, Freier SM, Bennett CF, Lollo B, Griffey R (2004) MicroRNA-143 regulates adipocyte differentiation. J Biol Chem 279:52361–52365.

Esau CC, Monia BP (2007) Therapeutic potential for microRNAs. Adv Drug Delivery Rev 59:101–114.

Farh KK-H, Grimson A, Jan C, Lewis BP, Johnston WK, Lim LP, Burge CB, Bartel DP (2005) The widespread impact of mammalian microRNAs on mRNA repression and evolution. Science 310:1817–1821.

Fedorov Y, Anderson EM, Birmingham A, Reynolds A, Karpilow J, Robinson K, Leake D, Marshall WS, Khvorova A. (2006) Off-target effects by siRNA can induce toxic phenotype. RNA 12:1188–1196.

Fernandez-Velasco M, Goren N, Benito G, Blanco-Rivero J, Bosca L, Delgado C (2003) Regional distribution of hyperpolarization-activated current I_f and hyperpolarization-activated cyclic nucleotide-gated channel mRNA expression in ventricular cells from control and hypertrophied rat hearts. J Physiol 553:395–405.

Fjose A, Drivenes O (2006) RNAi and microRNAs: From animal models to disease therapy. Birth Defects Res C Embryo Today 78:150–171.

Fujita S, Ito T, Mizutani T, Minoguchi S, Yamamichi N, Sakurai K, Iba H (2008) miR-21 Gene expression triggered by AP-1 is sustained through a double-negative feedback mechanism. J Mol Biol 378:492–504.

Furnari FB, Adams MD, Pagano JS (1993) Unconventional processing of the 30 termini of the Epstein-Barr virus DNA polymerase mRNA. Proc Natl Acad Sci USA 90:378–382.

Gao H, Xiao J, Sun Q, Lin H, Bai Y, Yang L, Yang B, Wang H, Wang Z (2006) A single decoy oligodeoxynucleotides targeting multiple oncoproteins produces strong anticancer effects. Mol Pharmacol 70:1621–1629.

Georges M, Coppieters W, Charlier C (2007) Polymorphic miRNA-mediated gene regulation: Contribution to phenotypic variation and disease. Curr Opin Genet Dev 17:166–176.

Gatignol A, Lainé S, Clerzius G (2005) Dual role of TRBP in HIV replication and RNA interference: Viral diversion of a cellular pathway or evasion from antiviral immunity? Retrovirology 2:65.

Ghuran AV, Camm AJ (2001) Ischemic heart disease presenting as arrhythmias. Br Med Bull 59:193–210.

Gillies JK, Lorimer IA (2007) Regulation of p27Kip1 by miRNA 221/222 in glioblastoma. Cell Cycle 6:2005–2009.

Gillet R, Felden B (2001) Emerging views on tmRNA-mediated protein tagging and ribosome rescue. Mol Microbiol 42:879–85.

Giraldez AJ, Cinalli RM, Glasner ME, Enright AJ, Thomson JM, Baskerville S, Hammond SM, Bartel DP, Schier AF (2005) MicroRNAs regulate brain morphogenesis in zebrafish. Science 308:833–838.

Gironella M, Seux M, Xie MJ, Cano C, Tomasini R, Gommeaux J, Garcia S, Nowak J, Yeung ML, Jeang KT, Chaix A, Fazli L, Motoo Y, Wang Q, Rocchi P, Russo A, Gleave M, Dagorn JC, Iovanna JL, Carrier A, Pébusque MJ, Dusetti NJ (2007) Tumor protein 53-induced nuclear protein 1 expression is repressed by miR-155, and its restoration inhibits pancreatic tumor development. Proc Natl Acad Sci USA 104:16170–16175.

Gottwein E, Cullen BR (2008) Viral and cellular microRNAs as determinants of viral pathogenesis and immunity. Cell Host Microbe 3:375–387.

Gottwein E, Mukherjee N, Sachse C, Frenzel C, Majoros WH, Chi JT, Braich R, Manoharan M, Soutschek J, Ohler U, Cullen BR (2007) A viral microRNA functions as an ortholog of cellular miR-155. Nature 450:1096–1099.

Gregory RI, Yan KP, Amuthan G, Chendrimada T, Doratotaj B, Cooch N, Shiekhattar R (2004) The Microprocessor complex mediates the genesis of microRNAs. Nature 432:235–240.

Grey F, Antoniewicz A, Allen E, Saugstad J, McShea A, Carrington JC, Nelson J (2005) Identification and characterization of human cytomegalovirus-encoded microRNAs. J Virol 79:12095–12099.

Gribaldo S, Brochier-Armanet C (2006) The origin and evolution of Archaea: A state of the art. Philos Trans R Soc Lond B Biol Sci 361:1007–1022.

Grimm D, Streetz KL, Jopling CL, Storm TA, Pandey K, Davis CR et al (2006) Fatality in mice due to oversaturation of cellular microRNA/short hairpin RNA pathways. Nature 441:537–541.

Grishok A, Pasquinelli AE, Conte D, Li N, Parrish S, Ha I, Baillie DL, Fire A, Ruvkun G, Mello CC (2001) Genes and mechanisms related to RNA interference regulate expression of the small temporal RNAs that control *C.*elegans developmental timing. Cell 106:23–34.

Grun D, Wang YL, Langenberger D, Gunsalus KC, Rajewsky N (2005) microRNAs target predictions across seven Drosophila species and comparison to mammalian targets. PLoS Comput Biol 1:e13.

Grundhoff A, Sullivan CS, Ganem D (2006) A combined computational and microarray-based approach identifies novel microRNAs encoded by human gamma-herpesviruses. RNA 12:733–750.

Gunawardane LS, Saito K, Nishida KM, Miyoshi K, Kawamura Y, Nagami T, Siomi H, Siomi MC (2007) A slicer-mediated mechanism for repeat-associated siRNA 5' end formation in Drosophila. Science 315:1587–1590.

Gupta A, Gartner JJ, Sethupathy P, Hatzigeorgiou AG, Fraser NW (2006) Anti-apoptotic function of a microRNA encoded by the HSV-1 latency-associated transcript. Nature 442:82–85.

Haasch D, Chen YW, Reilly RM, Chiou XG, Koterski S, Smith ML, Kroeger P, McWeeny K, Halbert DN, Mollison KW, Djuric SW, Trevillyan JM (2002) T cell activation induces a noncoding RNA transcript sensitive to inhibition by immunosuppressant drugs and encoded by the proto-oncogene, BIC. Cell Immunol 217:78–86.

Haley B, Zamore PD (2004) Kinetic analysis of the RNAi enzyme complex. Nat Struct Mol Biol 11, 599–606.

Hammond SM (2006) MicroRNAs as oncogenes. Curr Opin Genet Dev 16:4–9.

Han H, Long H, Wang H, Wang J, Zhang Y, Yang B, Wang Z (2004) Cellular remodeling of apoptosis in response to transient oxidative insult in rat ventricular cell line H9c2: A critical role of the mitochondria death pathway. Am J Physiol 286:H2169–H2182.

Han H, Wang H, Long H, Nattel S, Wang Z (2001) Oxidative preconditioning and apoptosis in L-cells: Roles of protein kinase B and mitogen-activated protein kinases. J Biol Chem 276:26357–26364.

Han J, Lee Y, Yeom KH, Nam JW, Heo I, Rhee JK, Sohn SY, Cho Y, Zhang BT, Kim VN (2006) Molecular basis for the recognition of primary microRNAs by the Drosha-DGCR8 complex. Cell 125:887–901.

Harris TA, Yamakuchi M, Ferlito M, Mendell JT, Lowenstein CJ (2008) MicroRNA-126 regulates endothelial expression of vascular cell adhesion molecule 1. Proc Natl Acad Sci USA 105:1516–1521.

Hatfield SD, Shcherbata HR, Fischer KA, Nakahara K, Carthew RW, Ruohola-Baker H (2005) Stem cell division is regulated by the microRNA pathway. Nature 435:974–978.

Hayashita Y, Osada H, Tatematsu Y, Yamada H, Yanagisawa K, Tomida S, Yatabe Y, Kawahara K, Sekido Y, Takahashi T (2005) A polycistronic microRNAs cluster, miR-17–92, is overexpressed in human lung cancers and enhances cell proliferation. Cancer Res 65:9628–9632.

Hayashi K, Chuva de Sousa Lopes SM, Kaneda M, Tang F, Hajkova P, Lao K, O'Carroll D, Das PP, Tarakhovsky A, Miska EA, Surani MA (2008) MicroRNA biogenesis is required for mouse primordial germ cell development and spermatogenesis. PLoS ONE 3:e1738.

He H, Jazdzewski K, Li W, Liyanarachchi S, Nagy R, Volinia S, Calin GA, Liu CG, Franssila K, Suster S, Kloos RT, Croce CM, de la Chapelle A (2005a) The role of microRNA genes in papillary thyroid carcinoma. Proc Natl Acad Sci USA 102:19075–19080.

He L, Thomson JM, Hemann MT, Hernando-Monge E, Mu D, Goodson S, Powers S, Cordon-Cardo C, Lowe SW, Hannon GJ, Hammond SM (2005b) A microRNA polycistron as a potential human oncogene. Nature 435:828–833.

Hino K, Tsuchiya K, Fukao T, Kiga K, Okamoto R, Kanai T, Watanabe M (2008) Inducible expression of microRNA-194 is regulated by HNF-1alpha during intestinal epithelial cell differentiation. RNA 14:1433–1442.

Hoheisel JD (2006) Microarray technology: Beyond transcript profiling and genotype analysis. Nat Rev Genet 7:200–210.

Horwich MD, Li C Matranga C, Vagin V, Farley G, Wang P, Zamore PD (2007) The *Drosophila* RNA methyltransferase, DmHen1, modifies germline piRNAs and single-stranded siRNAs in RISC. Current Biology 17:1265–1272.

Hu Z, Chen J, Tian T, Zhou X, Gu H, Xu L, Zeng Y, Miao R, Jin G, Ma H, Chen Y, Shen H (2008) Genetic variants of miRNA sequences and non-small cell lung cancer survival. J Clin Invest 118:2600–2608.

Humphreys DT, Westman BJ, Martin DI, Preiss T (2005) MicroRNAs control translation initiation by inhibiting eukaryotic initiation factor 4E/cap and poly(A) tail function. Proc Natl Acad Sci USA 102:16961–16966.

Hutvágner G, Zamore PD (2002) A microRNA in a multiple-turnover RNAi enzyme complex. Science 297:2056–2060.

Iorio MV, Ferracin M, Liu CG, Veronese A, Spizzo R, Sabbioni S, Magri E, Pedriali M, Fabbri M, Campiglio M, Ménard S, Palazzo JP, Rosenberg A, Musiani P, Volinia S, Nenci I, Calin GA, Querzoli P, Negrini M, Croce CM (2005) MicroRNA gene expression deregulation in human breast cancer. Cancer Res 65:7065–7070.

Ivey KN, Muth A, Arnold J, King FW, Yeh RF, Fish JE, Hsiao EC, Schwartz RJ, Conklin BR, Bernstein HS, Srivastava D (2008) MicroRNA regulation of cell lineages in mouse and human embryonic stem cells. Cell Stem Cell 2:219–229.

Izumiya Y, Kim S, Izumi Y, Yoshida K, Yoshiyama M, Matsuzawa A, Ichijo H, Iwao H (2003) Apoptosis signal-regulating kinase 1 plays a pivotal role in angiotensin II-induced cardiac hypertrophy and remodeling. Circ Res 93:874–883.

Jackson AL, Burchard J, Schelter J, Chau BN, Cleary M, Lim L, Linsley PS (2006) Widespread siRNA 'off-target' transcript silencing mediated by seed region sequence complementarity. RNA 12:1179–1187.

Jackson AL, Linsley PS (2004) Noise amidst the silence: Off-target effects of siRNAs? Trends Genet 20:521–524.

Jackson AL, Linsley PS (2003) Expression profiling reveals off-target gene regulation by RNAi. Nat Biotechnol 21:635–637.

Jackson RJ, Standart N (2007) How do microRNAs regulate gene expression? Sci STKE 23:243–249.

Jay C, Nemunaitis J, Chen P, Fulgham P, Tong AW (2007) miRNA profiling for diagnosis and prognosis of human cancer. DNA Cell Biol 26:293–300.

Jazdzewski K, Murray EL, Franssila K, Jarzab B, Schoenberg DR, de la Chapelle A (2008) Common SNP in pre-miR-146a decreases mature miR expression and predisposes to papillary thyroid carcinoma. Proc Natl Acad Sci USA 105:7269–7274.

Ji R, Cheng Y, Yue J, Yang J, Liu X, Chen H, Dean DB, Zhang C (2007) MicroRNA expression signature and antisense-mediated depletion reveal an essential role of MicroRNA in vascular neointimal lesion formation. Circ Res 100:1579–1588.

Jin P, Zarnescu DC, Ceman S, Nakamoto M, Mowrey J, Jongens TA, Nelson DL, Moses K, Warren ST (2004) Biochemical and genetic interaction between the fragile X mental retardation protein and the microRNA pathway. Nat Neurosci 7:113–117.

Johnson SM, Grosshans H, Shingara J, Byrom M, Jarvis R, Cheng A, Labourier E, Reinert KL, Brown D, Slack FJ (2005) RAS is regulated by the let-7 microRNA family. Cell 120:635–647.

Jopling CL, Yi M, Lancaster AM, Lemon SM, Sarnow P (2005) Modulation of hepatitis C virus RNA abundance by a liver-specific MicroRNA. Science 309:1577–1581.

Jost N, Virag L, Bitay M, Takacs J, Lengyel C, Biliczki P, Nagy Z, Bogats G, Lathrop DA, Papp JG, Varro A (2005) Restricting excessive cardiac action potential and QT prolongation: A vital role for I_{Ks} in human ventricular muscle. Circulation 112: 1392–1399.

Khvorova A, Reynolds A, Jayasena SD (2003) Functional siRNAs and miRNAs exhibit strand bias. Cell 115:209–216.

Kim VN (2005) MicroRNA biogenesis: Coordinated cropping and dicing. Nat Rev Mol Cell Biol 6:376–385.

Kim J, Krichevsky A, Grad Y, Hayes GD, Kosik KS, Church GM, Ruvkun G (2004) Identification of many microRNAs that copurify with polyribosomes in mammalian neurons. Proc Natl Acad Sci USA 101:360–365.

Kirchhoff F, Greenough TC, Brettler DB, Sullivan JL, Desrosiers RC (1995) Brief report: Absence of intact nef sequences in a long-term survivor with nonprogressive HIV-1 infection. N Engl J Med 332:228–232.

Kiss T (2001) Small nucleolar RNA-guided post-transcriptional modification of cellular RNAs. EMBO J 20:3617–3622.

Kloosterman WP, Plasterk RHA (2006) The diverse functions of microRNAs in animal development and disease. Dev Cell 11:441–450.

Knight SW, Bass BL (2001) A role for the RNase III enzyme DCR-1 in RNA interference and germ line development in *Caenorhabditis elegans*. Science 293:2269–2271.

Kosik KS, Krichevsky AM (2005) The Elegance of the MicroRNAs: A Neuronal Perspective. Neuron 47:779–782.

Krek A, Grün D, Poy MN, Wolf R, Rosenberg L, Epstein EJ, MacMenamin P, da Piedade I, Gunsalus KC, Stoffel M, Rajewsky N (2005) Combinatorial microRNA target predictions. Nat Genet 37:495–500.

Krichevsky AM, King KS, Donahue CP, Khrapko K, Kosik KS (2003) A microRNA array reveals extensive regulation of microRNAs during brain development. RNA 9:1274–1281.

Krützfeldt J, Rajewsky N, Braich R, Rajeev KG, Tuschl T, Manoharan M, Stoffel M (2005) Silencing of microRNAs in vivo with 'antagomirs'. Nature 438:685–689.

Krützfeldt J, Stoffel M (2006) MicroRNAs: A new class of regulatory genes affecting metabolism. Cell Metab 4:9–12.

Kuehbacher A, Urbich C, Zeiher AM, Dimmeler S (2007) Role of Dicer and Drosha for endothelial microRNA expression and angiogenesis. Circ Res 101:59–68.

Kumar A (2007) The silent defense: Micro-RNA directed defense against HIV-1 replication. Retrovirology 4:26.

Kumar MS, Lu J, Mercer KL, Golub TR, Jacks T (2007) Impaired microRNA processing enhances cellular transformation and tumorigenesis. Nat Genet 39:673–677.

Kuehbacher A, Urbich C, Zeiher AM, Dimmeler S (2007) Role of Dicer and Drosha for endothelial microRNA expression and angiogenesis. Circ Res 101:59–68.

Kuwabara T, Hsieh J, Nakashima K, Taira K, Gage FH (2004) A small modulatory dsRNA specifies the fate of adult neural stem cells. Cell 116:779–793.

Kwon C, Han Z, Olson EN, Srivastava D (2005) MicroRNA1 influences cardiac differentiation in Drosophila and regulates Notch signaling. Proc Natl Acad Sci USA 102:18986–18991.

Lagos-Quintana M, Rauhut R, Lendeckel W, Tuschl T (2001) Identification of novel genes coding for small expressed RNAs. Science 294:853–858.

Lagos-Quintana M, Rauhut R, Yalcin A, Meyer J, Lendeckel W, Tuschl T (2002) Identification of tissue-specific microRNAs from mouse. Curr Biol 12:735–739.

Lakshmipathy U, Love B, Goff LA, Jörnsten R, Graichen R, Hart RP, Chesnut JD (2007) MicroRNA expression pattern of undifferentiated and differentiated human embryonic stem cells. Stem Cells Dev 16:1003–1016.

Landgraf P, Rusu M, Sheridan R, Sewer A, Iovino N, Aravin A, Pfeffer S, Rice A, Kamphorst AO, Landthaler M, Lin C, Socci ND, Hermida L, Fulci V, Chiaretti S, Foà R, Schliwka J, Fuchs U, Novosel A, Müller RU, Schermer B, Bissels U, Inman J, Phan Q, Chien M, Weir DB, Choksi R, De Vita G, Frezzetti D, Trompeter HI, Hornung V, Teng G, Hartmann G, Palkovits M, Di Lauro R, Wernet P, Macino G, Rogler CE, Nagle JW, Ju J, Papavasiliou FN, Benzing T, Lichter P, Tam W, Brownstein MJ, Bosio A, Borkhardt A, Russo JJ, Sander C, Zavolan M, Tuschl T (2007) A mammalian microRNA expression atlas based on small RNA library sequencing. Cell 129:1401–1414.

Landi D, Gemignani F, Naccarati A, Pardini B, Vodicka P, Vodickova L, Novotny J, Försti A, Hemminki K, Canzian F, Landi S (2008) Polymorphisms within micro-RNA-binding sites and risk of sporadic colorectal cancer. Carcinogenesis 29:579–584.

Landthaler M, Yalcin A, Tuschl T (2004) The human DiGeorge syndrome critical region gene 8 and Its D. melanogaster homolog are required for miRNA biogenesis. Curr Biol 14:2162–2167.

Lau N, Lim L, Weinstein E, Bartel DP (2001) An abundant class of tiny RNAs with probable regulatory roles in *Caenorhabditis elegans*. Science 294:858–862.

Lecellier CH, Dunoyer P, Arar K, Lehmann-Che J, Eyquem S, Himber C, Saïb A, Voinnet O (2005) A cellular microRNA mediates antiviral defense in human cells. Science 308:557–560.

Lee RC, Ambros V (2001) An extensive class of small RNAs in *Caenorhabditis elegans*. Science 294:862–864.

Lee RC, Feinbaum RL, Ambros V (1993) The *C. elegans* heterochronic gene lin-4 encodes small RNAs with antisense complementarity to lin-14. Cell 75:843–854.

Lee Y, Hur I, Park SY, Kim YK, Suh MR, Kim VN (2006) The role of PACT in theRNA silencing pathway. EMBO J 25:522–532.

Lee Y, Jeon K, Lee JT, Kim S, Kim VN (2002a) MicroRNA maturation: Stepwise processing and subcellular localization. EMBO J 21:4663–4670.

Lee TI, Rinaldi NJ, Robert F, Odom DT, Bar-Joseph Z, Gerber GK, Hannett NM, Harbison CT, Thompson CM, Simon I, Zeitlinger J, Jennings EG, Murray HL, Gordon DB, Ren B, Wyrick JJ, Tagne JB, Volkert TL, Fraenkel E, Gifford DK, Young RA (2002b) Transcriptional regulatory networks in Saccharomyces cerevisiae. Science 298:799–804.

Lewis BP, Shih IH, Jones-Rhoades MW, Bartel DP, Burge CB (2003) Prediction of mammalian microRNA targets. Cell 115:787–798.

Lewis BP, Burge CB, Bartel DP (2005) Conserved seed pairing, often flanked by adenosines, indicates that thousands of human genes are microRNA targets. Cell 120:15–20.

Li M, Jones-Rhoades MW, Lau NC, Bartel DP, Rougvie AE (2005) Regulatory mutations of mir-48, a C. elegans let-7 family MicroRNA, cause developmental timing defects. Dev Cell 9:415–422.

Lim LP, Lau NC, Garrett-Engele P, Grimson A, Schelter JM, Castle J, Bartel DP, Linsley PS, Johnson JM (2005) Microarray analysis shows that some microRNAs downregulate large numbers of target mRNAs. Nature 433:769–773.

Lin SL, Chang SJ, Ying SY (2006) Transgene-like animal models using intronic microRNAs. Methods Mol Biol 342:321–334.

Lin X, Ruan X, Anderson MG, McDowell JA, Kroeger PE, Fesik SW, Shen Y (2005) siRNA-mediated off-target gene silencing triggered by a 7 nt complementation. Nucleic Acids Res 33:4527–4535.

Liu DW, Antzelevitch C (1995) Characteristics of the delayed rectifier current (I_{Kr} and I_{Ks}) in canine ventricular epicardial, midmyocardial, and endocardial myocytes. A weaker I_{Ks} contributes to the longer action potential of the M cell. Circ Res 76:351–365.

Liu Y, Mochizuki K, Gorovsky MA (2004) Histone H3 lysine 9 methylation is required for DNA elimination in developing macronuclei in Tetrahymena. Proc Natl Acad Sci USA 101:1679–1684.

Liu N, Williams AH, Kim Y, McAnally J, Bezprozvannaya S, Sutherland LB, Richardson JA, Bassel-Duby R, Olson EN (2007) An intragenic MEF2-dependent enhancer directs muscle-specific expression of microRNAs 1 and 133. Proc Natl Acad Sci USA 104:20844–20849.

Llave C, Kasschau KD, Rector MA, Carrington JC (2002) Endogenous and silencing-associated small RNAs in plants. Plant Cell 14:1605–1619.

Lu S, Cullen BR (2004) Adenovirus VA1 noncoding RNA can inhibit small interfering RNA and MicroRNA biogenesis. J Virol 78:12868–12876.

Lu J, Getz G, Miska EA, Alvarez-Saavedra E, Lamb J, Peck D, Sweet-Cordero A, Ebert BL, Mak RH, Ferrando AA, Downing JR, Jacks T, Horvitz HR, Golub TR (2005) MicroRNA expression profiles classify human cancers. Nature 435:834–838.

Lu Y, Thomson JM, Wong HY, Hammond SM, Hogan BL (2007) Transgenic over-expression of the microRNA miR-17–92 cluster promotes proliferation and inhibits differentiation of lung epithelial progenitor cells. Dev Biol 310:442–453.

Lu Y, Zhou J, Xu C, Lin H, Xiao J, Wang Z, Yang B (2008) AK/STAT and PI3K/AKT pathways form a mutual transactivation loop and afford lasting resistance to oxidative stress-induced apoptosis in cardiomyocytes. Cell Physiol Biochem 21:305–314.

Lund E, Guttinger S, Calado A, Dahlberg JE, Kutay U (2004) Nuclear export of microRNA precursors. Science 303:95–98.

Luo X, Lin H, Lu Y, Li B, Xiao J, Yang B, Wang Z (2007) Transcriptional activation by stimulating protein 1 and post-transcriptional repression by muscle-specific microRNAs of I_{Ks}-encoding genes and potential implications in regional heterogeneity of their expressions. J Cell Physiol 212:358–367.

Luo X, Lin H, Pan Z, Xiao J, Zhang Y, Lu Y, Yang B, Wang Z (2008) Overexpression of Sp1 and downregulation of miR-1/miR-133 activates re-expression of pacemaker channel genes HCN2 and HCN4 in hypertrophic heart. J Biol Chem 283:20045–20052.

Lustig AJ (1999) Crisis intervention: The role of telomerase. Proc Natl Acad Sci USA 96:3339–3341.

Ma JB, Yuan YR, Meister G, Pei Y, Tuschl T, Patel DJ (2005) Structural basis for 50-endspecific recognition of guide RNA by the A. fulgidus Piwi protein. Nature 434:666–670.

Mack GS (2007) MicroRNA gets down to business. Nat Technol 25:631–638.

McKinsey TA, Olson EN (2005) Toward transcriptional therapies for the failing heart: Chemical screens to modulate genes. J Clin Invest 115:538–546.

Mayr C, Hemann MT, Bartel DP (2007) Disrupting the pairing between let-7 and Hmga2 enhances oncogenic transformation. Science 315:1576–1579.

Meister G, Tuschl T (2004) Mechanisms of gene silencing by double-stranded RNA. Nature 431:343–349.

Mello CC, Conte Jr D (2004) Revealing the world of RNA interference. Nature 431:338–342.

Meng F, Henson R, Wehbe-Janek H, Smith H, Ueno Y, Patel T (2007) The MicroRNA let-7a modulates interleukin-6-dependent STAT-3 survival signaling in malignant human cholangiocytes. J Biol Chem 282:8256–8264.

Mersey BD, Jin P, Danner DJ (2005) Human microRNA (miR29b) expression controls the amount of branched chain a-ketoacid dehydrogenase complex in a cell, Hum Mol Genet 14:3371–3377.

Michael MZ, O' Connor SM, van Holst Pellekaan NG, Young GP, James RJ (2003) Reduced accumulation of specific microRNAs in colorectal neoplasia. Mol Cancer Res 1:882–891.

Miranda KC, Huynh T, Tay Y, Ang YS, Tam WL, Thomson AM, Lim B, Rigoutsos I (2006) A pattern-based method for the identification of MicroRNA binding sites and their corresponding heteroduplexes. Cell 126:1203–1217.

Mishra PJ, Humeniuk R, Mishra PJ, Longo-Sorbello GS, Banerjee D, Bertino JR (2007) A miR-24 microRNA binding-site polymorphism in dihydrofolate reductase gene leads to methotrexate resistance. Proc Natl Acad Sci USA 104:13513–13518.

Mishra PJ, Mishra PJ, Banerjee D, Bertino JR (2008) MiRSNPs or MiR-polymorphisms, new players in microRNA mediated regulation of the cell: Introducing microRNA pharmacogenomics. Cell Cycle 7:853–858.

Miska EA, Alvarez-Saavedra E, Townsend M, Yoshii A, Sestan N, Rakic P, Constantine-Paton M, Horvitz HR (2004) Constantine-Paton, H.R. Horitz, Microarray analysis of microRNA expression in the developing mammalian brain. Genome Biol 5:R68.

Mochizuki K, Gorovsky MA (2004a) Conjugationspecific small RNAs in Tetrahymena have predicted properties of scan (scn) RNAs involved in genome rearrangement. Genes Dev 18:2068–2073.

Mochizuki K, Gorovsky MA (2004b) Small RNAs in genome rearrangement in Tetrahymena. Curr Opin Genet Dev 14:181–187.

Mochizuki K, Fine NA, Fujisawa T, Gorovsky MA (2002) Analysis of a piwi-related gene implicates small RNAs in genome rearrangement in tetrahymena. Cell 110:689–699.

Moffat J, Sabatini DM (2006) Building mammalian signalling pathways with RNAi screens. Nat Rev Mol Cell Biol 7:177–187.

Morita S, Horii T, Kimura M, Goto Y, Ochiya T, Hatada I (2007) One Argonaute family member, Eif2c2 (Ago2), is essential for development and appears not to be involved in DNA methylation. Genomics 89:687–696.

Motsch N, Pfuhl T, Mrazek J, Barth S, Grässer FA (2007) Epstein-Barr virus-encoded latent membrane protein 1 (LMP1) induces the expression of the cellular microRNA miR-146a. RNA Biol 4:131–137.

Mourelatos Z, Dostie J, Paushkin S, Sharma A, Charroux B, Abel L, Rappsilber J, Mann M, Dreyfuss G (2002) miRNPs: A novel class of Ribonucleo proteins containing numerous microRNAs. Genes Dev 16:720–728.

Mudhasani R, Zhu Z, Hutvagner G, Eischen CM, Lyle S, Hall LL, Lawrence JB, Imbalzano AN, Jones SN (2008) Loss of miRNA biogenesis induces p19Arf-p53 signaling and senescence in primary cells. J Cell Biol 181:1055–1063.

Murchison EP, Stein P, Xuan Z, Pan H, Zhang MQ, Schultz RM, Hannon GJ (2007) Critical roles for Dicer in the female germline. Genes Dev 21:682–693.

Nair V, Zavolan M (2006) Virus-encoded microRNAs: Novel regulators of gene expression. Trends Microbiol 14:169–715.

Nelson PT, Baldwin DA, Kloosterman WP, Kauppinen S, Plasterk RH, Mourelatos Z (2006) RAKE and LNA-ISH reveal microRNA expression and localization in archival human brain. RNA 12:187–191.

Nilsen TW (2007) Mechanisms of microRNA-mediated gene regulation in animal cells. Trends Genet 23:243–249.

Niu Z, Li A, Zhang SX, Schwartz RJ (2007) Serum response factor micromanaging cardiogenesis. Curr Opin Cell Biol 19:618–627.

O'Carroll D, Mecklenbrauker I, Das PP, Santana A, Koenig U, Enright AJ, Miska EA, Tarakhovsky A (2007) A Slicer-independent role for Argonaute 2 in hematopoiesis and the microRNA pathway. Genes Dev 21:1999–2004.

O'Connell RM, Taganov KD, Boldin MP, Cheng G, Baltimore D (2007) MicroRNA-155 is induced during the macrophage inflammatory response. Proc Natl Acad Sci USA 104:1604–1609.

Okamura K, Hagen JW, Duan H, Tyler DM, Lai EC (2007) The mirtron pathway generates microRNA-class regulatory RNAs in Drosophila. Cell 130:89–100.

Okamura K, Ishizuka A, Siomi H, Siomi MC (2004) Distinct roles for Argonaute proteins in small RNA-directed RNA cleavage pathways. Genes Dev 18:1655–1666.

Omoto S, Fujii YR (2005) Regulation of human immunodeficiency virus 1 transcription by nef microRNA. J Gen Virol 86:751–755.

Omoto S, Ito M, Tsutsumi Y, Ichikawa Y, Okuyama H, Brisibe EA, Saksena NK, Fujii YR (2004) HIV-1 nef suppression by virally encoded microRNA. Retrovirology 1:44.

Otsuka M, Jing Q, Georgel P, New L, Chen J, Mols J, Kang YJ, Jiang Z, Du X, Cook R, Das SC, Pattnaik AK, Beutler B, Han J (2007) Hypersusceptibility to vesicular stomatitis virus infection in Dicer1-deficient mice is due to impaired miR24 and miR93 expression. Immunity 27:123–134.

Otsuka M, Zheng M, Hayashi M, Lee JD, Yoshino O, Lin S, Han J (2008) Impaired microRNA processing causes corpus luteum insufficiency and infertility in mice. J Clin Invest 118:1944–1954.

Pannucci JA, Haas ES, Hall TA, Harris JK, Brown JW (1999) RNase P RNAs from some Archaea are catalytically active. Proc Natl Acad Sci USA 96:7803–7808.

Park MY, Wu G, Gonzalez-Sulser A, Vaucheret H, Poethig RS (2005) Nuclear processing and export of microRNAs in Arabidopsis. Proc Natl Acad Sci USA 102:3691–3696.

Parker JS, Roe SM, Barford D (2005) Structural insights into mRNA recognition from a PIWI domain-siRNA guide complex. Nature 434:663–666.

Pasquinelli AE, Reinhart BJ, Slack F, Martindale MQ, Kuroda MI, Maller B, Hayward DC, Ball EE, Degnan B, Muller P, Spring J, Srinivasan A, Fishman M, Finnerty J, Corbo J, Levine M, Leahy P, Davidson E, Ruvkun G (2000) Conservation of the sequence and temporal expression of let-7 heterochronic regulatory RNA. Nature 408:86–89.

Pedersen IM, Cheng G, Wieland S, Volinia S, Croce CM, Chisari FV, David M (2007) Interferon modulation of cellular microRNAs as an antiviral mechanism. Nature 449:919–922.

Peragine A, Yoshikawa M, Wu G, Albrecht HL, Poethig RS (2004) SGS3 and SGS2/SDE1/RDR6 are required for juvenile development and the production of transacting siRNAs in Arabidopsis. Genes Dev 18:2368–2379.

Perreault J, Perreault J-P, Boire G (2007) Ro-associated Y RNAs in metazoans: Evolution and diversification. Mol Biol Evol 24:1678–1689.

Pfeffer S, Sewer A, Lagos-Quintana M, Sheridan R, Sander C, Grässer FA, van Dyk LF, Ho CK, Shuman S, Chien M, Russo JJ, Ju J, Randall G, Lindenbach BD, Rice CM, Simon V, Ho DD, Zavolan M, Tuschl T (2005) Identification of microRNAs of the herpesvirus family. Nat Methods 2:269–276.

Pfeffer S, Zavolan M, Grässer FA, Chien M, Russo JJ, Ju J, John B, Enright AJ, Marks D, Sander C, Tuschl T (2004) Identification of virus-encoded microRNAs. Science 304:734–736.

Pillai RS, Bhattacharyya SN, Artus CG, Zoller T, Cougot N, Basyuk E, Bertrand E, Filipowicz W (2005) Inhibition of translational initiation by Let-7 MicroRNA in human cells. Science 309:1573–1576.

Pillai RS, Bhattacharyya SN, Filipowicz W (2007) Repression of protein synthesis by miRNAs: How many mechanisms? Trends Cell Biol 17:18–126.

Plaisance V, Abderrahmani A, Perret-Menoud V, Jacquemin P, Lemaigre F, Regazzi R (2006) MicroRNA-9 controls the expression of Granuphilin/Slp4 and the secretory response of insulin-producing cells. J Biol Chem 281:26932–26942.

Poy MN, Spranger M, Stoffel M (2007) microRNAs and the regulation of glucose and lipid metabolism. Diabetes Obes Metab Suppl 2:67–73.

Rajewsky N, Socci ND (2004) Computational identification of microRNA targets. Dev Biol 267:529–535.

Randall G, Panis M, Cooper JD, Tellinghuisen TL, Sukhodolets KE, Pfeffer S, Landthaler M, Landgraf P, Kan S, Lindenbach BD, Chien M, Weir DB, Russo JJ, Ju J, Brownstein MJ, Sheridan R, Sander C, Zavolan M, Tuschl T, Rice CM (2007) Cellular cofactors affecting hepatitis C virus infection and replication. Proc Natl Acad Sci USA 104:12884–12889.

Raveche ES, Salerno E, Scaglione BJ, Manohar V, Abbasi F, Lin YC, Fredrickson T, Landgraf P, Ramachandra S, Huppi K, Toro JR, Zenger VE, Metcalf RA, Marti GE (2007) Abnormal microRNA-16 locus with synteny to human 13q14 linked to CLL in NZB mice. Blood 109:5079–5086.

Raver-Shapira N, Marciano E, Meiri E, Spector Y, Rosenfeld N, Moskovits N, Bentwich Z, Oren M (2007) Transcriptional activation of miR-34a contributes to p53-mediated apoptosis. Mol Cell 26:731–743.

Reinhart BJ, Slack FJ, Basson M, Pasquinelli AE, Bettinger JC, Rougvie AE, Horvitz HR, Ruvkun G (2000) The 21-nucleotide let-7 RNA regulates developmental timing in Caenorhabditis elegans. Nature 403:901–906.

Ricke DO, Wang S, Cai R, Cohen D (2006) Genomic approaches to drug discovery. Curr Opin Chem Biol 10:303–308.

Rodríguez M, Lucchesi BR, Schaper J (2002) Apoptosis in myocardial infarction. Ann Med 34:470–479.

Rodriguez A, Vigorito E, Clare S, Warren MV, Couttet P, Soond DR, van Dongen S, Grocock RJ, Das PP, Miska EA, Vetrie D, Okkenhaug K, Enright AJ, Dougan G, Turner M, Bradley A (2007) Requirement of bic/microRNA-155 for normal immune function. Science 316:608–611.

Ruby JG, Jan C, Bartel DP (2007) Intronic microRNA precursors that bypass Drosha processing. Nature 448:83–86.

Ruvkun G (2001) Molecular biology. Glimpses of a tiny RNA world. Science 294:797–799.

Saetrom P, Snøve O Jr, Rossi JJ (2007) Epigenetics and microRNAs. Pediatr Res 61:17R–23R.

Saito Y, Liang G, Egger G, Friedman JM, Chuang JC, Coetzee GA, Jones PA (2006) Specific activation of microRNA-127 with downregulation of the proto-oncogene BCL6 by chromatin-modifying drugs in human cancer cells. Cancer Cell 9:435–443.

Sall A, Liu Z, Zhang HM, Yuan J, Lim T, Su Y, Yang D (2008) MicroRNAs-based therapeutic strategy for virally induced diseases. Curr Drug Discov Technol 5:49–58.

Salvi R, Garbuglia AR, Di Caro A, Pulciani S, Montella F, Benedetto A (1998) Grossly defective nef gene sequences in a human immunodeficiency virus type 1-seropositive long-term nonprogressor. J Virol 72:3646–3657.

Samols MA, Hu J, Skalsky RL, Renne R (2005) Cloning and identification of a microRNA cluster within the latency-associated region of Kaposi's sarcoma-associated herpesvirus. J Virol 79:9301–9305.

Saunders MA, Liang H, Li WH (2007) Human polymorphism at microRNAs and microRNA target sites. Proc Natl Acad Sci USA 104:3300–3305.

Sayed D, Hong C, Chen IY, Lypowy J, Abdellatif M (2007) MicroRNAs play an essential role in the development of cardiac hypertrophy. Circ Res 100:416–424.

Scaria V, Hariharan M, Pillai B, Maiti S, Brahmachari SK (2007) Host-virus genome interactions: Macro roles for microRNAs. Cell Microbiol 9:2784–2794.

Scaria V, Jadhav V (2007) microRNAs in viral oncogenesis. Retrovirology 4:82.

Schat KA, Calnek BW (1978) Protection against Marek's disease-derived tumor transplants by the nononcogenic SB-1 strain of Marek's disease virus. Infect Immun 22:225–232.

Schratt GM, Tuebing F, Nigh EA, Kane CG, Sabatini ME, Kiebler M, Greenberg ME (2006) A brain-specific microRNA regulates dendritic spine development. Nature 439:283–289.

Schwarz DS, Hutvagner G, Du T, Xu Z, Aronin N, Zamore PD (2003) Asymmetry in the assembly of the RNAi enzyme complex. Cell 115:199–208.

Scott GK, Mattie MD, Berger CE, Benz SC, Benz CC (2006) Rapid alteration of microRNA levels by histone deacetylase inhibition. Cancer Res 66:1277–1281.

Sempere LF, Freemantle S, Pitha-Rowe I, Moss E, Dmitrovsky E, Ambros V (2004) Expression profiling of mammalian microRNAs uncovers a subset of brain-expressed microRNAs with possible roles in murine and human neuronal differentiation, Genome Biol 5:R13.

Sethupathy P, Borel C, Gagnebin M, Grant GR, Deutsch S, Elton TS, Hatzigeorgiou AG, Antonarakis SE (2007) Human microRNA-155 on chromosome 21 differentially interacts with its polymorphic target in the AGTR1 3' untranslated region: A mechanism for functional single-nucleotide polymorphisms related to phenotypes. Am J Hum Genet 81:405–413.

Shen J, Ambrosone CB, DiCioccio RA, Odunsi K, Lele SB, Zhao H (2008) A functional polymorphism in the miR-146a gene and age of familial breast/ovarian cancer diagnosis. Carcinogenesis 29:1963–1966.

Shyu AB, Wilkinson MF, van Hoof A (2008) Messenger RNA regulation: To translate or to degrade. EMBO J 27:471–481.

Si ML, Zhu S, Wu H, Lu Z, Wu F, Mo YY (2007) miR-21-mediated tumor growth. Oncogene 26:2799–1803.

Silva JM, Hammond SM, Hannon GJ (2002) RNA interference: A promising approach to antiviral therapy? Trends Mol Med 8:505–508.

Singh SK (2007) miRNAs: From neurogeneration to neurodegeneration. Pharmacogenomics 8:971–978.

Skalsky RL, Samols MA, Plaisance KB, Boss IW, Riva A, Lopez MC, Baker HV, Renne R (2007) Kaposi's sarcoma-associated herpesvirus encodes an ortholog of miR-155. J Virol 81:12836–12845.

Slack FJ, Weidhaas JB (2006) MicroRNAs as a potential magic bullet in cancer. Future Oncol 2:73–82.

Smirnova L, Gräfe A, Seiler A, Schumacher S, Nitsch R, Wulczyn FG (2005) Regulation of miRNA expression during neural cell specification. Eur J Neurosci 21:1469–1477.

Soifer HS, Rossi JJ, Saetrom P (2007) MicroRNAs in disease and potential therapeutic applications. Mol Ther 15:2070–2079.

Song JJ, Smith SK, Hannon GJ, Joshua-Tor L (2004) Crystal structure of Argonaute and its implications for RISC slicer activity. Science 305:1434–1437.

Srivastava D, Thomas T, Lin Q, Kirby ML, Brown D, Olson EN (1997) Regulation of cardiac mesodermal and neural crest development by the bHLH transcription factor, dHAND. Nat Genet 16:154–160.

Stenvang J, Kauppinen S (2008) MicroRNAs as targets for antisense-based therapeutics. Expert Opin Biol Ther 8:59–81.

Stilli D, Sgoifo A, Macchi E, Zaniboni M, De Iasio S, Cerbai E, Mugelli A, Lagrasta C, Olivetti G, Musso E (2001) Myocardial remodeling and arrhythmogenesis in moderate cardiac hypertrophy in rats. Am J Physiol 280:H142–H150.

Suárez Y, Fernández-Hernando C, Pober JS, Sessa WC (2007) Dicer dependent microRNAs regulate gene expression and functions in human endothelial cells. Circ Res 100:1164–1173.

Sullivan CS, Ganem D (2005) MicroRNAs and viral infection. Mol Cell 20:3–7.

Sullivan CS, Grundhoff AT, Tevethia S, Pipas JM, Ganem D (2005) SV40-encoded microRNAs regulate viral gene expression and reduce susceptibility to cytotoxic T cells. Nature 435:682–686.

Szentadrassy N, Banyasz T, Biro T, Szabo G, Toth BI, Magyar J, Lazar J, Varro A, Kovacs L, Nanasi PP (2005) Apico-basal inhomogeneity in distribution of ion channels in canine and human ventricular myocardium. Cardiovasc Res 65:851–860.

Taganov KD, Boldin MP, Chang KJ, Baltimore D (2006) NF-kappaB-dependent induction of microRNA miR-146, an inhibitor targeted to signaling proteins of innate immune responses. Proc Natl Acad Sci USA 103:12481–12486.

Takamizawa J, Konishi H, Yanagisawa K, Tomida S, Osada H, Endoh H, Harano T, Yatabe Y, Nagino M, Nimura Y, Mitsudomi T, Takahashi T (2004) Reduced expression of the let-7 microRNAs in human lung cancers in association with shortened postoperative survival. Cancer Res 64:3753–3756.

Tam W (2001) Identification and characterization of human BIC, a gene on chromosome 21 that encodes a noncoding RNA. Gene 274:157–167.

Tam W, Ben-Yehuda D, Hayward WS (1997) Bic, a novel gene activated by proviral insertions in avian leukosis virus-induced lymphomas, is likely to function through its noncoding RNA. Mol Cell Biol 17:1490–1502.

Tanzer A, Stadler PF (2004) Molecular evolution of a microRNA cluster. J Mol Biol 339:327–335.

Tarasov V, Jung P, Verdoodt B, Lodygin D, Epanchintsev A, Menssen A, Meister G, Hermeking H (2007) Differential regulation of microRNAs by p53 revealed by massively parallel sequencing: miR-34a is a p53 target that induces apoptosis and G1-arrest. Cell Cycle 6:1586–1593.

Tatsuguchi M, Seok HY, Callis TE, Thomson JM, Chen JF, Newman M, Rojas M, Hammond SM, Wang DZ (2007) Expression of microRNAs is dynamically regulated during cardiomyocyte hypertrophy. J Mol Cell Cardiol 42:1137–1141.

Tay Y, Zhang J, Thomson AM, Lim B, Rigoutsos I (2008) MicroRNAs to Nanog, Oct4 and Sox2 coding regions modulate embryonic stem cell differentiation. Nature 455:1124–1128.

Tazawa H, Tsuchiya N, Izumiya M, Nakagama H (2007) Tumor-suppressive miR-34a induces senescence-like growth arrest through modulation of the E2F pathway in human colon cancer cells. Proc Natl Acad Sci USA 104:15472–15477.

Tea BS, Dam TV, Moreau P, Hamet P, deBlois D (1999) Apoptosis during regression of cardiac hypertrophy in spontaneously hypertensive rats. Temporal regulation and spatial heterogeneity. Hypertension 34:229–235.

Teiger E, Than VD, Richard L, Wisnewsky C, Tea BS, Gaboury L, Tremblay J, Schwartz K, Hamet P (1996) Apoptosis in pressure overload-induced heart hypertrophy in the rat. J Clin Invest 97:2891–2897.

Teleman AA, Maitra S, Cohen SM (2005) Drosophila lacking microRNA miR-278 are defective in energy homeostasis. Genes Dev 20:417–422.

Thai TH, Calado DP, Casola S, Ansel KM, Xiao C, Xue Y, Murphy A, Frendewey D, Valenzuela D, Kutok JL, Schmidt-Supprian M, Rajewsky N, Yancopoulos G, Rao A, Rajewsky K (2007) Regulation of the germinal center response by microRNA-155. Science 316:604–608.

Thore S, Mayer C, Sauter C, Weeks S, Suck D (2003) Crystal Structures of the Pyrococcus abyssi Sm Core and Its Complex with RNA. J Biol Chem 278:1239–1247.

Thum T, Catalucci D, Bauersachs J (2008) MicroRNAs: Novel regulators in cardiac development and disease. Cardiovasc Res 79:562–570.

Thum T, Galuppo P, Wolf C, Fiedler J, Kneitz S, van Laake LW, Doevendans PA, Mummery CL, Borlak J, Haverich A, Gross C, Engelhardt S, Ertl G, Bauersachs J (2007) MicroRNAs in the human heart: A clue to fetal gene reprogramming in heart failure. Circulation 116:258–267.

Triboulet R, Mari B, Lin YL, Chable-Bessia C, Bennasser Y, Lebrigand K, Cardinaud B, Maurin T, Barbry P, Baillat V, Reynes J, Corbeau P, Jeang KT, Benkirane M (2007) Suppression of microRNA-silencing pathway by HIV-1 during virus replication. Science 315:1579–1582.

Tuddenham L, Wheeler G, Ntounia-Fousara S, Waters J, Hajihosseini MK, Clark I, Dalmay T (2006) The cartilage specific microRNA-140 targets histone deacetylase 4 in mouse cells. FEBS Lett 580:4214–4217.

van den Berg A, Kroesen BJ, Kooistra K, de Jong D, Briggs J, Blokzijl T, Jacobs S, Kluiver J, Diepstra A, Maggio E, Poppema S (2003) High expression of B-cell receptor inducible gene BIC in all subtypes of Hodgkin lymphoma. Genes Chromosomes Cancer 37:20–28.

van Rooij E, Sutherland LB, Liu N, Williams AH, McAnally J, Gerard RD, Richardson JA, Olson EN (2006) A signature pattern of stress-responsive microRNAs that can evoke cardiac hypertrophy and heart failure. Proc Natl Acad Sci USA 103:18255–18260.

van Rooij E, Sutherland LB, Qi X, Richardson JA, Hill J, Olson EN (2007) Control of stress-dependent cardiac growth and gene expression by a microRNA. Science 316:575–579.

Vasudevan S, Tong Y, Steitz JA (2007) Switching from repression to activation: MicroRNAs can up-regulate translation. Science 318:1931–1934.

Vazquez F, Vaucheret H (2004) Endogenous trans-acting siRNAs regulate the accumulation of Arabidopsis mRNAs. Mol Cell 16:1–13.

Verduyn SC, Vos MA, van der Zande J, van der Hulst FF, Wellens HJ (1997) Role of interventricular dispersion of repolarization in acquired torsade-de-pointes arrhythmias: Reversal by magnesium. Cardiovasc Res 34:453–463.

Vidal L, Blagden S, Attard G, de Bono J (2005) Making sense of antisense. Eur J Cancer 41:2812–2818.

Vigorito E, Perks KL, Abreu-Goodger C, Bunting S, Xiang Z, Kohlhaas S, Das PP, Miska EA, Rodriguez A, Bradley A, Smith KG, Rada C, Enright AJ, Toellner KM, Maclennan IC, Turner M (2007) microRNA-155 regulates the generation of immunoglobulin class-switched plasma cells. Immunity 27:847–859.

Volinia S, Calin GA, Liu CG, Ambs S, Cimmino A, Petrocca F, Visone R, Iorio M, Roldo C, Ferracin M, Prueitt RL, Yanaihara N, Lanza G, Scarpa A, Vecchione A, Negrini M, Harris CC, Croce CM (2006) A microRNA expression signature of human solid tumors defines cancer gene targets. Proc Natl Acad Sci USA 103:2257–2261.

Wang B, Love TM, Call ME, Doench JG, Novina CD (2006) Recapitulation of short RNA-directed translational gene silencing in vitro. Mol Cell 22:553–560.

Wang Z, Feng J, Shi H, Pond A, Nerbonne JM, Nattel S (1999) The potential molecular basis of different physiological properties of transient outward K^+ current in rabbit and human hearts. Circ Res 84:551–561.

Wang Z, Luo X, Lu Y, Yang B (2008) miRNAs at the heart of the matter. J Mol Med 86:772–783.

Wang Z, Yue L, White M, Pelletier G, Nattel S (1998) Differential expression of inward rectifier potassium channel mRNA in human atrium versus ventricle and in normal versus failing hearts. Circulation 98:2422–2428.

Welch C, Chen Y, Stallings RL (2007) MicroRNA-34a functions as a potential tumor suppressor by inducing apoptosis in neuroblastoma cells. Oncogene 26:5017–5022.

Wienholds E, Koudijs MJ, van Eeden FJ, Cuppen E, Plasterk RH (2003) The microRNA-producing enzyme Dicer1 is essential for zebrafish development. Nat Genet 35:217–218.

Wightman B, Ha I, Ruvkun G (1993) Posttranscriptional regulation of the heterochronic gene lin-14 by lin-4 mediates temporal pattern formation in *C. elegans*. Cell 75:855–562.

Wilfred BR, Wang WX, Nelson PT (2007) Energizing miRNA research: A review of the role of miRNAs in lipid metabolism, with a prediction that miR-103/107 regulates human metabolic pathways. Mol Genet Metab 91:209–217.

Wurdinger T, Costa FF (2007) Molecular therapy in the microRNA era. Pharmacogenomics J 7:297–304.

Woodhams MD, Stadler PF, Penny D, Collins LJ (2007) RNase MRP and the RNA processing cascade in the eukaryotic ancestor. BMC Evol Biol 7:S13.

Xiao J, Luo X, Lin H, Xu C, Gao H, Wang H, Yang0 B, Wang Z (2007) MicroRNA miR-133 represses HERG K^+ channel expression contributing to QT prolongation in diabetic hearts. J Biol Chem 282:12363–12367.

Xu C, Lu Y, Lin H, Xiao J, Wang H, Luo X, Li B, Yang B, Wang Z (2007) The muscle-specific microRNAs miR-1 and miR-133 produce opposing effects on apoptosis via targeting HSP60/HSP70 and caspase-9 in cardiomyocytes. J Cell Sci 120:3045–3052.

Xu P, Vernooy SY, Guo M, Hay BA (2003) The Drosophila microRNA Mir-14 suppresses cell death and is required for normal fat metabolism. Curr Biol 13:790–795.

Xu T, Zhu Y, Wei QK, Yuan Y, Zhou F, Ge YY, Yang JR, Su H, Zhuang SM (2008) A functional polymorphism in the miR-146a gene is associated with the risk for hepatocellular carcinoma. Carcinogenesis 29:2126–2131.

Yanaihara N, Caplen N, Bowman E, Seike M, Kumamoto K, Yi M, Stephens RM, Okamoto A, Yokota J, Tanaka T, Calin GA, Liu CG, Croce CM, Harris CC (2006) Unique microRNA molecular profiles in lung cancer diagnosis and prognosis. Cancer Cell 9:189–198.

Yang B, Lin H, Xiao J, Lu Y, Luo X, Li B, Wang H, Chen G, Wang Z (2007) The muscle-specific microRNA miR-1 causes cardiac arrhythmias by targeting GJA1 and KCNJ2 genes. Nat Med 13:486–491.

Yang B, Lu Y, Wang Z (2008) Control of cardiac excitability by microRNAs. Cardiovasc Res 79:571–580.

Yang M, Mattes J (2008) Discovery, biology and therapeutic potential of RNA interference, microRNA and antagomirs. Pharmacol Ther 117:94–104.

Yekta S, Shih IH, Bartel DP (2004) MicroRNA-directed cleavage of HOXB8 mRNA. Science 304:594–596.

Yi R, Qin Y, Macara IG, Cullen BR (2003) Exportin-5 mediates the nuclear export of pre-microRNAs and short hairpin RNAs. Genes Dev 17:3011–3016.

Yin Q, McBride J, Fewell C, Lacey M, Wang X, Lin Z, Cameron J, Flemington EK (2008) MicroRNA-155 is an Epstein-Barr virus-induced gene that modulates Epstein-Barr virus-regulated gene expression pathways. J Virol 82:5295–5306.

Ying SY, Lin SL (2005) Intronic microRNAs. Biochem Biophys Res Commun 326:515–520.

Yu B, Yang Z, Li J, Minakhina S, Yang M, Padgett RW, Steward R, Chen X (2005) Methylation as a crucial step in plant microRNAs biogenesis. Science 307:932–935.

Yu Z, Li Z, Jolicoeur N, Zhang L, Fortin Y, Wang E, Wu M, Shen SH (2007) Aberrant allele frequencies of the SNPs located in microRNA target sites are potentially associated with human cancers. Nucleic Acids Res 35:4535–4541.

Yu XY, Song YH, Geng YJ, Lin QX, Shan ZX, Lin SG, Li Y (2008) Glucose induces apoptosis of cardiomyocytes via microRNA-1 and IGF-1. Biochem Biophys Res Commun 376:548–552.
Zhang C (2008) MicroRNomics: A newly emerging approach for disease biology. Physiol Genomics 33:139–147.
Zhang Y, Lu Y, Wang N, Lin H, Pan Z, Gao X, Zhang F, Zhang Y, Xiao J, Shan H, Luo X, Chen G, Qiao G, Wang Z, Yang B (2009) Control of experimental atrial fibrillation by microRNA-328. Science (in revision)
Zhang Y, Xiao J, Wang H, Luo X, Wang J, Villeneuve LR, Zhang H, Bai Y, Yang B, Wang Z (2006) Restoring depressed HERG K^+ channel function as a mechanism for insulin treatment of the abnormal QT prolongation and the associated arrhythmias in diabetic rabbits. Am J Physiol 291:1446–1455.
Zhao Y, Ransom JF, Li A, Vedantham V, von Drehle M, Muth AN, Tsuchihashi T, McManus MT, Schwartz RJ, Srivastava D (2007) Dysregulation of cardiogenesis, cardiac conduction, and cell cycle in mice lacking miRNA-1–2. Cell 129:303–317.
Zhao Y, Samal E, Srivastava D (2005) Serum response factor regulates a muscle specific microRNA that targets Hand2 during cardiogenesis. Nature 436:214–220.
Zhu S, Si ML, Wu H, Mo YY (2007) MicroRNA-21 targets the tumor suppressor gene tropomyosin 1 (TPM1). J Biol Chem 282:14328–14336.
Zidar N, Jera J, Maja J, Dusan S (2007) Caspases in myocardial infarction. Adv Clin Chem 44:1–33.

Chapter 2
miRNA Interference Technologies: An Overview

Abstract The miRNA pathways are highly responsive to interventions of any kind, being excellent candidates for pharmacological manipulation. Mature miRNAs belong to a category of non-coding small RNAs. They are very different from traditional messenger, ribosomal, or transfer RNA in their biogenesis, structures and functions. The technologies used to study traditional RNAs in basic research and to manipulate traditional RNAs for therapeutic purpose, for the most part, would be ineffective for miRNAs. New technologies for studying miRNA are therefore in need and this demand is being met by the scientific community at a breakneck pace. In the past few years, from the discovery of miRNAs in humans, numerous new technologies have been developed to clone, profile, visualize, quantify and manipulate miRNAs. This book centers on the technologies for manipulating miRNAs.

An overview of the concept of microRNA interference (miRNAi) and another three related concepts and a brief introduction to the miRNAi technologies is given in this chapter, to help readers better understand the technologies. The applications of these technologies to miRNA research and the potential for gene therapy of related diseases are also summarized and speculated. Detailed descriptions of these technologies are provided in the following chapters.

2.1 New Concepts of miRNAi Technologies

2.1.1 "miRNAi", A New Concept

RNA interference (RNAi) is a well-known strategy for gene silencing; this strategy takes the advantages of the capability of small double-stranded RNA molecules (siRNAs) to bind RNA-induced silencing complex (RISC) on the one hand and to bind target genes (mRNAs) on the other (Xia et al. 2002; Golden et al. 2008; Pushparaj et al. 2008). Through such dual interactions, siRNAs elicit a powerful knockdown of gene expression by degrading their target mRNAs. Two key

Z. Wang, *MicroRNA Interference Technologies*,
DOI: 10.1007/978-3-642-00489-6_2, © Springer-Verlag Berlin Heidelberg 2009

characteristics of the RNAi strategy are that the only target of RNAi is mRNAs and that the only outcome of RNAi is silencing of mRNAs. In other words, the RNAi strategy uses siRNAs to interfere directly with mRNAs (mostly protein-coding genes) to silence gene expression.

Taking the same concept, I propose a new concept of microRNA interference (Wang et al. 2008). miRNAi manipulates the function, stability, biogenesis, or expression of miRNAs and as such it indirectly interferes with the expression of protein-coding mRNAs. This concept is based on the thoughts outlined below. The fundamental mechanism of miRNA regulation of gene expression is miRNA:mRNA interaction or binding. A key to interfere with miRNA actions is to disrupt or to facilitate the miRNA:mRNA interaction. In order to achieve this aim, one can either manipulate or to facilitate miRNAs or mRNAs to alter the miRNA:mRNA interaction. For miRNAs, one can either mimic miRNA actions to enhance the miRNA:mRNA interaction or to inhibit miRNAs to break the miRNA:mRNA interaction. Additionally, one can also manipulate mRNA to interrupt the miRNA:mRNA interaction.

I further propose to call the strategies for interfering with miRNAs miRNAi technologies. miRNAi technologies can be categorized based on the following six different perspectives.

1. According to the mechanisms of actions, miRNAi technologies can be divided into two major strategies: miRNA-targeting and targeting-miRNAs strategies.

The "miRNA-targeting strategy of miRNAi" refers to the approaches producing "gain-of-function" of miRNAs to enhance gene targeting so as to alter the gene expression and cellular function (Chaps. 3–6).

The "targeting-miRNA strategy of miRNAi" refers to the approaches leading to "loss-of-function" of miRNAs via inhibiting miRNA expression and/or action to alter the gene expression and cellular function (Chaps. 7–13).

2. In terms of the outcome of actions, miRNAi technologies can be grouped into "miRNA-gain-of-function" (enhancement of miRNA function) and "miRNA-loss-of-function" (inhibition of miRNA expression, biogenesis, or function).

The "miRNA-gain-of-function" approach is in general achieved by overexpression of the endogenous miRNAs and forced expression of exogenous miRNAs, which results in enhanced miRNA targeting (Chaps. 3–6).

The "miRNA-loss-of-function" approach can be conferred through targeting-miRNAs by knockdown or knockout of miRNA expression and inhibition of miRNA action (Chaps. 7–13).

3. With respect to the target miRNAs, miRNAi technologies can be miRNA-specific or non-miRNA-specific.

The miRNA-specific miRNAi technologies interfere only with miRNA function and through this, they produce derepression of target protein-coding genes. They belong to the "targeting-miRNA" strategy being miRNA-specific but non-gene-specific.

The non-miRNA-specific miRNAi technologies interfere with the biogenesis of miRNAs such as inhibition of Dicer, affecting the levels of the whole population of miRNAs but not a particular miRNA. These approaches are therefore neither miRNA-specific nor gene-specific.

4. As to the target genes of miRNAs, miRNAi technologies comprise non-gene-specific and gene-specific approaches, which can be either "gain-of-function" of miRNAs or "loss-of-function" of miRNAs.

The non-gene-specific miRNAi technologies direct their actions to miRNAs without interaction with the target genes of the miRNAs. Such actions are miRNA-specific but not gene-specific given that a miRNA can target multiple protein-coding genes. Most of the currently available miRNAi technologies belong to non-gene-specific approaches.

The gene-specific miRNAi acts on the target gene of a given miRNA but not on the miRNA per se. This is achieved by enhancing or removing the actions of miRNAs but leaving miRNAs intact.

5. From the perspective of target protein-coding genes (mRNAs), miRNAi technologies may result in either "mRNA-gain-of-function" (derepression of genes) or "mRNA-loss-of-function" (enhancement or establishment of repression).

The "targeting-miRNA" strategy causes "mRNA-gain-of-function" by relieving the repressive actions induced by the targeted miRNAs. The "mRNA-gain-of-function" miRNAi technologies could be gene-specific or non-gene-specific but they must be miRNA-specific.

The "miRNA-targeting" strategy causes "mRNA-loss-of-function" effects by enhancing the repressive actions of the miRNAs. The "mRNA-loss-of-function" miRNAi technologies could be gene-specific or non-gene-specific but they must be miRNA-specific.

6. Taking the mode of actions into consideration, miRNAi technologies can act by creating AMO:miRNA interaction (AMO: anti-miRNA antisense oligomer) or by creating ASO:mRNA interaction (ASO: anti-mRNA antisense oligomer) to disrupt normal miRNA:mRNA interaction.

The AMO:miRNA-interacting miRNAi technologies suppress miRNA function using AMO approaches. They are all miRNA-specific and non-gene-specific, belonging to the targeting-miRNAs strategies and the miRNA-loss-of-function and mRNA-gain-of-function class.

The ASO:mRNA-interacting miRNAi technologies interrupt miRNA function using ASO approaches. They belong to the targeting-miRNAs strategies and the miRNA-loss-of-function and mRNA-gain-of-function class.

Based on above classifications, some miRNAs are both miRNA- and gene-specific, some are miRNA-specific but non-gene-specific and others are neither miRNA-specific nor gene-specific. Table 2.1 summarizes the six miRNAi strategies. In the following chapters, each miRNAi technology will be specified for its classification.

Table 2.1 General Introduction to miRNAi

Mechanism of action	miRNA-targeting	producing "miRNA-gain-of-function" to enhance "mRNA-loss-of-function"
	Targeting-miRNAs	producing "miRNA-gain-of-function" to enhance "mRNA-loss-of-function"
Mode of action	AMO:miRNA interaction	anti-miRNA oligomer inhibiing miRNA level or activity
	ASO:mRNA interaction	anti-mRNA oligomer inhibiting miRNA function
Outcome of miRNA	miRNA-gain-of-function	enhancement of miRNA function
	miRNA-loss-of-function	inhibition of miRNA expression, biogenesis, or function
Outcome of target gene	mRNA-gain-of-function	causing derepression of protein-coding genes
	mRNA-loss-of-function	enhancing repression of protein-coding genes
miRNA specificity	miRNA-specific	targeting a specific miRNA or a specific group of miRNAs
	Non-miRNA-specific	targeting the whole population of miRNAs
Gene specificity	Non-gene-specific	a miRNA targeting multiple genes with its binding site
	Gene-specific	a miRNA targeting a particular gene

2.1.2 "miRNA as a Regulator of a Cellular Function", Second New Concept

As already mentioned in the previous chapter, with respect to its pathophysiological role, a miRNA can be viewed as a regulator of a particular cellular function or a particular cellular program, not of a single gene. I propose this concept based upon the following facts (Wang et al. 2008).

1. Each miRNA normally has multiple target genes on the one hand and each gene may be a target for multiple miRNAs on the other hand. Focusing on any particular one of the target genes may lead to incomplete understanding of function of a miRNA.
2. The target genes of a given miRNA may encode proteins that have different or even opposing functions (e.g., cell growth vs. cell death). Focusing on any particular one of the target genes could be misleading to our understanding of exact function of a miRNA. For example, the oncoprotein Bcl-2 and the tumor suppressor p53 are both the target genes of miR-150 and miR-214. If one looks at only Bcl-2, then one may conclude that both miR-214 and miR-150 are miRNA tumor suppressors; conversely, if one focuses on p53, then one will believe both miR-214 and miR-150 to be oncomiRs: two contradictory views of the same miRNAs.

3. The net outcome of integral actions of a miRNA on its multiple target genes defines the function of that miRNA. Following this idea, whether miR-214 and miR-150 are tumor suppressors or oncomiRs will depend upon which of the target genes p53 and Bcl-2 predominates under a particular cellular context. If the effects on p53 predominate, then miR-214 and miR-150 are oncomiRs but if it is the other way around, then miR-214 and miR-150 are tumor suppressors.

Therefore, a more reasonable, correct view of function of a miRNA is that a miRNA is a regulator of a cellular function or a cellular program.

This concept is important to keep in mind when designing or applying miRNAi technologies in one's studies.

2.1.3 "One-Drug, Multiple-Target", Third New Concept

Human diseases are mostly multifactorial and multistep processes. Targeting a single factor (molecule) may not be adequate and certainly not optimal in disease therapy, because single agents are limited by incomplete efficacy and dose-limiting adverse effects. If related factors are concomitantly attacked, better outcomes are expected and the combination pharmacotherapy has been developed based on this reasoning: a combination of two or more drugs or therapeutic agents given as a single treatment that produces improved therapeutic results. The "drug cocktail" therapy of AIDS is one example of such a strategy (Henkel 1999), and similar approaches have been used for a variety of other diseases, including cancers (Konlee 1998; Charpentier 2002; Ogihara 2003; Kumar, 2005; Lin et al. 2005; Nabholtz and Gligorov 2005). However, the current drug-cocktail therapy is costly and may involve a complicated treatment regimen, undesired drug-drug interactions and increased side effects (Konlee 1998). There is a need to develop a strategy to avoid these problems.

In 2006, I proposed the "One-Drug, Multiple-Target" concept: a single agent capable of acting on multiple selected key targets to treat a disease (Gao et al. 2006). One example of such a drug is amiodarone, a class III antiarrhythmic agent. While most of the specific ion channel blockers can increase mortality by producing *de nova* lethal arrhythmias, amiodarone, by targeting multiple ion channels without channel specificity, can produce beneficial antiarrhythmic effects devoid of an increase in the risk of mortality. However, with respect to its highly desirable actions, amiodarone is an unexpected, accidental pharmaceutical product, not from a purposed design; it is nearly impossible to confer to single compounds the ability to act on multiple target molecules with the traditional pharmaceutical approaches or the currently known antigene strategies. My laboratory developed a complex decoy oligonucleotides strategy based on the "One-Drug, Multiple-Target" concept, which has demonstrated its superiority to inhibit tumor growth. In theory, the "One-Drug, Multiple-Target" strategy mimics the well-known drug cocktail therapy of AIDS. Nevertheless, this "One-Drug, Multiple-Target" strategy may be devoid of the weaknesses of the drug cocktail therapy, without involving complicated treatment regimens, undesirable drug-drug interactions and increased side effects. It opens the door to one-drug, multiple-target intervention, providing promising prototypes of gene therapeutic agents for a wide range of human disease.

I introduce here this concept because several of the miRNAi technologies have been developed based on this concept. Indeed, considering the fact that a single miRNA has the potential to regulate multiple target genes and a single gene may be regulated by multiple miRNAs, it is often necessary to interfere with multiple miRNAs or multiple genes to acquire effective manipulation of gene expression and cellular function. The "One-Drug, Multiple-Target" strategy opens up new opportunities for creative and rational designs of a variety of combinations integrating varying miRNA-related elements for a therapeutic purpose and provides an exquisite tool for functional analysis of miRNAs in a gene controlling program. It can also be used as a simple and straightforward approach for studying any other biological process involving multiple factors, multiple genes and multiple signaling pathways.

2.1.4 "miRNA Seed Family", Another New Concept

We have introduced the "Seed Site" concept proposed by Lewis et al. (2003, 2005) in Sect. 1.2 to define the mechanism of target recognition and action of miRNAs (i.e., miRNA:mRNA interactions). Based on this concept, miRNAs possessing a same seed motif (5′-end 2–8 nts) should have the same repertoire of target genes thereby the same cellular function. This concept has indeed been verified by numerous experimental investigations.

For the sake of easiness and clarity in understanding the function of miRNAs, I propose to categorize miRNAs into families based on their function or seed motifs. According to this classification system, miRNAs with a same seed motif 5′-end 2–8 nts are grouped into the same miRNA seed family. Further, miRNAs carrying exactly the same seed motif are grouped into the same miRNA seed subfamily. For example, miR-17-5p and miR-20b have identical 5′-end 1–8 nts seed sequence CAAAGUGC; miR-520g and miR-520h have ACAAAGUG; miR-20a and miR-106b contains UAAAGUGC; miR-106a, AAAAGUGC; miR-93, miR-372 and miR-520a-e all have AAAGUGCU; miR-519b and miR-519c have AAAGUGCA. Intriguingly, if 7 of 8 nts in the seed motif base-pairing with target genes is sufficient to produce post-transcriptional repression (as already shown by an enormous volume of studies) (Lewis et al. 2003, 2005; Pillai et al. 2007), then these six seed motifs should all give the same cellular effects. Based on this view, I consider all these miRNAs as the members of one miRNA seed family while belonging to six different subfamilies.

This classification provides a guideline of pivotal importance for interfering with a cellular process involving gene expression regulation by a multi-member miRNA seed family. Enhancing or inhibiting any one of the members of a miRNA seed family may not be able to elicit efficient, thorough changes of gene expression and cellular function. In this case, manipulation of all members of a miRNA seed family

AGUGC Family

Subfamily A

```
5′-ACAAAGUGCUUCCCUUUAGAGUGU-3′    (has-miR-520g)
5′-ACAAAGUGCUUCCCUUUAGAGU-3′      (has-miR-520h)
```

Subfamily B

```
5′- CAAAGUGCUUACAGUGCAGGUAG-3′    (has-miR-17-5p)
5′- UAAAGUGCUUAUAGUGCAGGUAG-3′    (has-miR-20a)
5′- AAAAGUGCUUACAGUGCAGGUAG-3′    (has-miR-106a)
5′- UAAAGUGCUGACAGUGCAGAU-3′      (has-miR-106b)
5′- CAAAGUGCUGUUCGUGCAGGUAG-3′    (has-miR-93)
5′- CAAAGUGCUCAUAGUGCAGGUAG-3′    (has-miR-20b)
5′- CAAAGUGCCUCCCUUUAGAGUG-3′     (has-miR-519d)
```

Subfamily C

```
5′-  AAAGUGCAUCCUUUUAGAGUGU-3′    (has-miR-519a)
5′-  AAAGUGCAUCUUUUUAGAGGAU-3′    (has-miR-519c-3p)
5′-  AAAGUGCAUCCUUUUAGAGGUU-3′    (has-miR-519b-3p)
```

Subfamily D

```
5′-  AAAGUGCUUCUCUUUGGUGGGU-3′    (has-miR-519d-3p)
5′-  AAAGUGCUUCCCUUUGGACUGU-3′    (has-miR-520a)
5′-  AAAGUGCUUCCUUUUAGAGGG-3′     (has-miR-520b)
5′-  AAAGUGCUUCCUUUUAGAGGGU-3′    (has-miR-520c-3p)
5′-  AAAGUGCUUCCUUUUUGAGGG-3′     (has-miR-520e)
5′-  GAAGUGCUUCGAUUUUGGGGUGU-3′   (has-miR-373)
5′-  UAAGUGCUUCCAUGCUU-3′         (has-miR-302e)
5′-  AAAGUGCUGCGACAUUUGAGCGU-3′   (has-miR-372)
```

Subfamily E

```
5′-   AAGUGCUGUCAUAGCUGAGGUC-3′   (has-miR-512-3p)
5′-   AAGUGCUUCCUUUUAGAGGGUU-3′   (has-miR-520f)
```

Fig. 2.1 An example of miRNA seed family concept. The seed site of each listed miRNA is shown in red-bold-face and highlighted in yellow. The miRNAs bearing exactly the same seed sites (5′-end 2–7 nts) are grouped into a subfamily. Assuming that 6-nt base-pairing out of 7-nt seed site is sufficient to produce gene targeting actions, these five subfamilies labeled A–E are supposed to have same target genes and cellular functions and they are therefore placed into one same family

is definitely required to achieve a level with sufficient miRNA-promoting effects or anti-miRNA effects.

We have sorted out all miRNAs registered in miRBase by their seed motifs and are able to categorize these miRNAs into 498 seed families. For convenience, I designate these families according to their 4–7 nts (Fig. 2.1). Some families contain subfamilies with varying number of miRNAs and some currently contain only one member.

2.2 General Introduction to miRNAi Technologies

2.2.1 miRNA-Targeting Technologies

An array of technologies has been developed for achieving gain-of-function of miRNAs by forced expression or overexpression. These include Synthetic Canonical miRNA technology, miRNA Mimic technology, Multi-miRNA Hairpins technology, Multi-miRNA Mimics technology and miRNA Transgene technology.

The Synthetic Canonical MiRNA technology involves an application of synthetic miRNAs that are identical in sequence to their counterpart endogenous miRNAs (Luo et al. 2007, 2008; Yang et al. 2007; Xiao et al. 2007a).

The key for the miRNA Mimic technology is to generate non-naturally existing double-stranded RNA fragments that are able to produce miRNA-like actions: the post-transcriptional repression of gene expression or inhibition of protein translation (Xiao et al. 2007b). It is a gene-specific miRNA-targeting strategy.

The Multi-miRNA Hairpins technology uses a single artificial construct in the form of double-stranded RNA to produce multiple mature miRNAs once introduced into cells (Sun et al. 2006; Xia et al. 2006).

The Multi-miRNA Mimics technology incorporates multiple miRNA Mimic units for targeting different mRNAs into a single construct that is able to silence multiple targeted genes (Chen et al. 2008a).

MiRNA Transgene technology requires production of mice by incorporating a non-native segment of DNA containing a pre-miRNA-coding sequence of interest into mice's germline, which retains the ability to overexpress that miRNA in the transgenic mice (Zhang et al. 2009).

2.2.2 Targeting-miRNA Technologies

Targeting-miRNA technologies include Anti-miRNA Antisense Inhibitor Oligoribonucleotides (AMO) technology, Multiple-Target AMO technology (MT-AMO technology), miRNA Sponge technology, miRNA-Masking Antisense Oligonucleotides technology, Sponge miR-Mask technology and MiRNA Knockout technology.

The Anti-miRNA Antisense Inhibitor Oligoribonucleotides (AMOs) are single-stranded 2′-*O*-methyl-modified oligoribonucleotide fragments exactly antisense to their target miRNAs and can bind to their target miRNAs to inhibit their actions.

The Multiple-Target AMO technology is a modified AMO strategy, which allows designing single-stranded 2′-*O*-methyl-modified oligoribonucleotides carrying multiple antisense units that are engineered into a single fragment that is able to simultaneously silence multiple targeted miRNAs or multiple miRNA seed families (Lu et al. 2009).

The miRNA Sponge technology involves synthesis of RNAs containing multiple, tandem binding sites for miRNAs with the same seed sequence (a miRNA seed family) of interest and is able to target all members of that miRNA seed family (Ebert et al. 2007; Hammond 2007).

The MiRNA-Masking Antisense Oligonucleotides technology involves synthesis of single-stranded 2′-*O*-methyl-modified oligoribonucleotides that does not directly interact with its target miRNA but binds the binding site of that miRNA in the 3′ UTR of the target mRNA by a fully complementary mechanism (Xiao et al. 2007b). In this way, the miR-Mask covers up the access of its target miRNA to the binding site to derepress its target gene via blocking the action of its target miRNA. It is a gene-specific targeting-miRNA strategy.

The Sponge miR-Mask technology combines the principle of actions of the MiRNA Sponge and the miR-Mask technologies for targeting miRNAs (Chen et al. 2008b). It is a gene-specific targeting-miRNA strategy.

The miRNA or Dicer Knockout technology enables to generate targeted deletion of a specific miRNA or Dicer; the former is miRNA-specific but non-mRNA-specific and the latter is non-miRNA-, non-mRNA-specific.

These miRNAi technologies are summarized in Table 2.2 and the strategies of miRNAi are illustrated in Fig. 2.2.

2.3 miRNAi Technologies in Basic Research and Drug Design

miRNAs are universally expressed in mammalian cells, being involved in nearly every aspect of the life of organisms (Lagos-Quintana et al. 2002; Landgraf et al. 2007). The level of miRNAs in cells is dynamic depending on developmental stage, cell cycle, metabolic alteration and pathophysiological conditions. Both upregulation and downregulation of miRNAs have been frequently implicated in a variety of pathological conditions. Both gain-of-function and loss-of-function technologies are necessary tools for understanding miRNAs. The miRNA-Targeting strategy resulting in gain-of-function of miRNAs is an essential way for gain-of-knowledge about miRNA targets and functions. On the other hand, the Targeting-miRNA technologies leading to loss-of-function of miRNAs is also an indispensable strategy for miRNA research. These two approaches are mutually complementary for elucidating miRNA biology and pathophysiology. The miRNAi technologies have opened up new opportunities for creative and rational designs of a variety of combinations integrating varying nucleotide fragments for various purposes and providing exquisite tools for functional analysis related to identification and characterization of targets of miRNAs and their functions in a gene controlling program.

In addition, the miRNAi technologies offer the strategies and tools for designing new agents for gene therapy of human disease (Ambros et al. 2003; Lee et al. 2005; Bartel, 2004; Wang et al. 2008; Calin & Croce, 2006). These agents will possess a backbone structure in the form of oligoribonucleotides or oligodeoxyribonucleotides. Like other types of nucleic acids for gene therapy, such as antisense

Table 2.2 Summary of miRNAi technologies

Name of miRNAi Technology	Characteristics	Applications
miRNA-targeting miRNA-gain-of-function mRNA-loss-of-function	Enhancing miRNA action to enhance gene silencing	
Synthetic Canonical miRNA (SC-miRNA)	miRNA-specific Non-gene-specific	(1) transient miRNA-gain-of-function in miRNA research; (2) miRNA replacement therapy
miRNA Mimic (miR-Mimic)	Non-miRNA-specific Gene-specific	(1) transient miRNA-gain-of-function in a gene-specific manner for miRNA research; (2) miRNA replacement therapy
Multi-miRNA Hairpins	miRNA-specific Non-gene-specific	(1) transient miRNA-gain-of-function in miRNA research; (2) miRNA replacement therapy
Multi-miRNA Mimics (Multi-miR-Mimic)	Non-miRNA-specific Gene-specific	(1) transient miRNA-gain-of-function in a gene-specific manner for miRNA research; (2) miRNA replacement therapy
miRNA Transgene	miRNA-specific Non-gene-specific	(1) long-lasting miRNA-gain-of-function in miRNA research; (2) controllable or conditional overexpression of miRNA; (3) in vivo animal model; (4) miRNA replacement therapy
Targeting-miRNA miRNA-loss-of-function mRNA-gain-of-function	Knockdown or knockout miRNA to relieve gene silencing	
Anti-miRNA Antisense Oligonucleotides (AMO)	miRNA-specific Non-gene-specific	(1) To knockdown target miRNA for validating the miRNA targets & function; (2) To achieve upregulation of the cognate target protein; (3) To reverse the pathological process
Multiple-Target Anti-miRNA Antisense Oligonucleotides (MT-AMO)	miRNA-specific Non-gene-specific	(1) To knockdown multiple miRNAs from same or different seed families; (2) To achieve upregulation of multiple target proteins; (3) To reverse the pathological process
miRNA Sponge	miRNA-specific Non- gene-specific	(1) To knockdown multiple miRNAs from a same seed family; (2) To achieve upregulation of the cognate target proteins; (3) To reverse the pathological process

(*continued*)

Table 2.2 (Continued)

Name of miRNAi Technology	Characteristics	Applications
miRNA-Masking Antisense Oligonucleotides (miR-Mask)	miRNA-specific Gene-specific	(1) To block action of a miRNA without knocking down the miRNA; (2) To achieve upregulation of the cognate target protein; (3) To reverse the pathological process
Sponge miR-Mask	miRNA-specific Non-gene-specific	(1) To knockdown multiple miRNAs from a same seed family; (2) To achieve upregulation of the cognate target proteins; (3) To reverse the pathological process
miRNA Knockout (miR-KO)	miRNA-specific Non-gene-specific	(1) To acquire permanent removal of a target miRNA; (2) To allow for controllable or conditional miRNA silencing; (3) To allow for studying miRNA-loss-of-function in vivo whole-animal context
Dicer Inactivation	Non-miRNA-specific Non-gene-specific	(1) To achieve a global loss-of-function of literally all cellular miRNAs; (2) To allow for controllable or conditional miRNA silencing; (3) To allow for studying miRNA-loss-of-function in vivo whole-animal context

oligodeoxynucleotides (Alvarez-Salas 2008; Koizumi 2007), decoy oligodeoxynucleotides (Gao et al. 2006), siRNA (Alvarez-Salas 2008; Pushparaj et al. 2008), triplex-forming oligodeoxynucleotides (Mahato et al. 2005), aptamer (Kaur and Roy 2008) and DNAzyme (Benson et al. 2008), the oligomer fragments generated using miRNAi technologies can be chemically modified to improve stability and are constructed into plasmids for easier delivery into organisms to treat diseases.

The Synthetic Canonical miRNA technology is an indispensable and most essential approach for miRNA gain-of-function in the fundamental research of miRNAs and of miRNA-related biological processes. It has been usually used to force expression of miRNAs of interest to investigate the pathophysiological outcome of upregulation of the miRNAs. The approach can also be used to supplement the loss of the miRNAs under certain situations and to maintain the normal levels to alleviate the pathological conditions as a result of downregulation of those miRNAs. The Multi-miRNA Hairpins technology can be employed if more than one miRNAs need to be applied under certain conditions.

miRNA Mimics can be used when a particular protein-coding gene needs to be knocked down. miRNA Mimics and siRMAs are quite similar in terms of

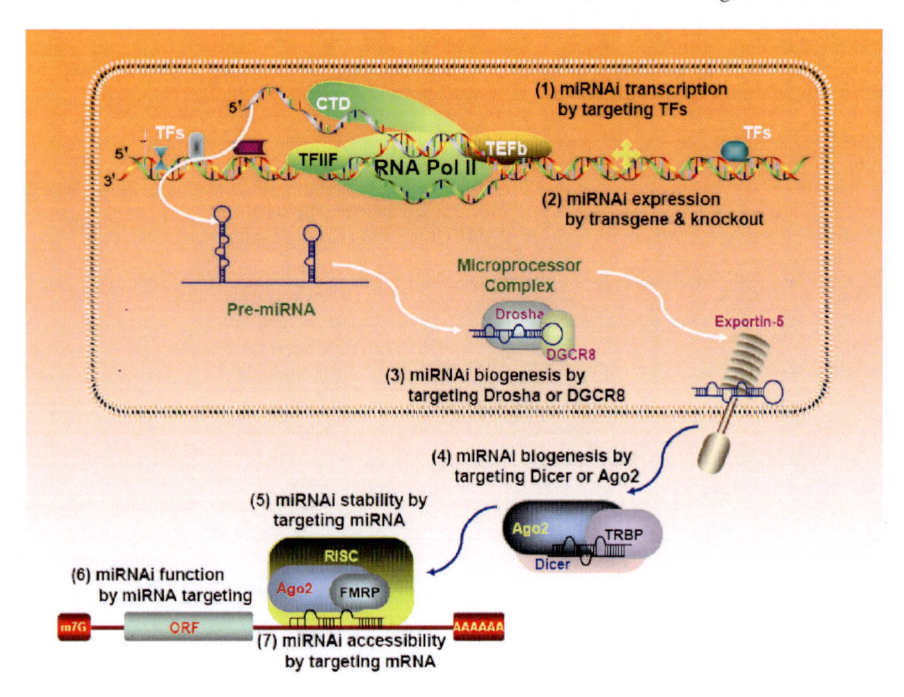

Fig. 2.2 Diagram illustrating the miRNAi strategies: **(1)** miRNAi can be directed to interfere with miRNA transcription by targeting transcriptional factors to either enhance or repress transcription. **(2)** miRNAi can change miRNA expression by creating Transgene or knockout for in vivo model studies. **(3)** and **(4)** miRNAi can disrupt miRNA biogenesis by targeting Drosha, DGCR8, Dicer and Ago2. **(5)** miRNAi can be directed to target the stability of miRNAs to knockdown miRNA level. **(6)** miRNAi can be used to change the function of miRNA by miRNA targeting (miRNA replacement). **(7)** miRNAi can block miRNA accessibility to its bining site in the target genes

their applications to basic research and their potential for disease treatment. The Multi-miRNA Mimics technology can be used as a simple and straightforward approach for studying the biological processes involving multiple protein-coding genes.

miRNA Transgene technology is an efficient approach to create overexpression of miRNAs of interest to study the role of miRNAs in in vivo conditions.

The AMO technology has been the most commonly and frequently used targeting-miRNA/loss-of-function approach to knock down miRNAs of interest. This method offers a miRNA-specific way to wipe out the function of the targeted miRNAs and can be used to bring back to the normal levels of miRNAs that are abnormally upregulated in the diseases states. The Multiple-Target AMO technology is designed as a simple and straightforward approach for studying the biological processes involving multiple miRNAs.

The MiRNA-Masking Antisense Oligonucleotides technology is a supplement to the AMO technique; while AMO is indispensable for studying the overall function of a miRNA, the miR-Mask might be more appropriate for studying the specific outcome of regulation of the target gene by the miRNA. The Sponge miR-Mask technology is an alternative for the applications requiring manipulation of multiple target protein-coding genes of miRNAs.

The miRNA Knockout technology aims to generate mouse lines with genetic ablation of specific miRNAs or targeted disruption of miRNA genes. This approach allows for investigations of miRNA function related to the development of particular biological processes and/or pathological conditions in an in vivo context and in a permanent manner.

Dicer Inactivation technology aims to disrupt the maturation of miRNAs by inhibiting Dicer function through knocking down or knocking out the Dicer gene. This approach has been widely used to study the global requirement of miRNAs for certain fundamental biological and pathological processes.

References

Alvarez-Salas LM (2008) Nucleic acids as therapeutic agents. Curr Top Med Chem 8:1379–1404.

Ambros V, Bartel B, Bartel DP, Burge CB, Carrington JC, Chen X, Dreyfuss G, Eddy SR, Griffiths-Jones S, Marshall M, Matzke M, Ruvkun G, Tuschl T (2003) A uniform system for microRNAs annotation. RNA 9:277–279.

Bartel DP (2004) MicroRNAs: Genomics, biogenesis, mechanism, and function. Cell 116:281–297.

Benson VL, Khachigian LM, Lowe HC (2008) DNAzymes and cardiovascular disease. Br J Pharmacol 154:741–748.

Calin GA, Croce CM (2006) MicroRNA-Cancer connection: The beginning of a new tale. Cancer Res 66:7390–7394.

Charpentier G (2002) Oral combination therapy for type 2 diabetes. Diabetes Metab Res Rev 18(Suppl 3):S70–S76.

Chen G, Lin H, Xiao J, Luo X, Wang Z (2008a) Multi-miRNA-Mimics strategy offers a powerful and diverse gain-of-function of miRNAs for gene silencing. Biotechniques.

Chen G, Lin H, Xiao J, Luo X, Wang H, Wang Z (2008b) Proapoptotic effect of miR-17-5p miRNA seed family in response to oxidative stress in cardiac myocytes. Nat Struct Mol Biol (in revision).

Ebert MS, Neilson JR, Sharp PA (2007) MicroRNA sponges: Competitive inhibitors of small RNAs in mammalian cells. Nat Methods 4:721–726.

Gao H, Xiao J, Sun Q, Lin H, Bai Y, Yang L, Yang B, Wang H, Wang Z (2006) A single decoy oligodeoxynucleotides targeting multiple oncoproteins produces strong anticancer effects. Mol Pharmacol 70:1621–1629.

Golden DE, Gerbasi VR, Sontheimer EJ (2008) An inside job for siRNAs. Mol Cell 31:309–312.

Hammond SM (2007) Soaking up small RNAs. Nat Methods 4:694–695.

Henkel J (1999) Attacking AIDS with a ‘cocktail’ therapy? FDA Consum 33:12–17.

Kaur G, Roy I (2008) Therapeutic applications of aptamers. Expert Opin Investig Drugs 17:43–60.

Koizumi M (2007) True antisense oligonucleotides with modified nucleotides restricted in the N conformation. Curr Top Med Chem 7:661–665.

Konlee M (1998) An evaluation of drug cocktail combinations for their immunological value in preventing/remitting opportunistic infections. Posit Health News 16:2–4.

Kumar P (2005) Combination treatment significantly enhances the efficacy of antitumor therapy by preferentially targeting angiogenesis. Lab Investig 85:756–767.

Lagos-Quintana M, Rauhut R, Yalcin A, Meyer J, Lendeckel W, Tuschl T (2002) Identification of tissue-specific microRNAs from mouse. Curr Biol 12:735–739.

Landgraf P, Rusu M, Sheridan R, Sewer A, Iovino N, Aravin A, Pfeffer S, Rice A, Kamphorst AO, Landthaler M, Lin C, Socci ND, Hermida L, Fulci V, Chiaretti S, Foà R, Schliwka J, Fuchs U, Novosel A, Müller RU, Schermer B, Bissels U, Inman J, Phan Q, Chien M, Weir DB, Choksi R, De Vita G, Frezzetti D, Trompeter HI, Hornung V, Teng G, Hartmann G, Palkovits M, Di Lauro R, Wernet P, Macino G, Rogler CE, Nagle JW, Ju J, Papavasiliou FN, Benzing T, Lichter P, Tam W, Brownstein MJ, Bosio A, Borkhardt A, Russo JJ, Sander C, Zavolan M, Tuschl T (2007) A mammalian microRNA expression atlas based on small RNA library sequencing. Cell 129:1401–1414.

Lee YS, Kim HK, Chung S, Kim KS, Dutta A (2005) Depletion of human micro-RNA miR-125b reveals that it is critical for the proliferation of differentiated cells but not for the down-regulation of putative targets during differentiation. J Biol Chem 280:16635–16641.

Lewis BP, Shih IH, Jones-Rhoades MW, Bartel DP, Burge CB (2003) Prediction of mammalian microRNA targets. Cell 115:787–798.

Lewis BP, Burge CB, Bartel DP (2005) Conserved seed pairing, often flanked by adenosines, indicates that thousands of human genes are microRNA targets. Cell 120:15–20.

Lin LM, Li BX, Xiao JB, Lin DH, Yang BF (2005) Synergistic effect of all-transretinoic acid and arsenic trioxide on growth inhibition and apoptosis in human hepatoma, breast cancer, and lung cancer cells in vitro. World J Gastroenterol 11:5633–5637.

Lu Y, Xiao J, Lin H, Luo X, Wang Z, Yang B (2009) Complex antisense inhibitors offer a superior approach for microRNA research and therapy. Nucleic Acids Res 37:e24.

Luo X, Lin H, Lu Y, Li B, Xiao J, Yang B, Wang Z (2007) Transcriptional activation by stimulating protein 1 and post-transcriptional repression by muscle-specific microRNAs of I_{Ks}-encoding genes and potential implications in regional heterogeneity of their expressions. J Cell Physiol 212:358–367.

Luo X, Lin H, Pan Z, Xiao J, Zhang Y, Lu Y, Yang B, Wang Z (2008) Overexpression of Sp1 and downregulation of miR-1/miR-133 activates re-expression of pacemaker channel genes HCN2 and HCN4 in hypertrophic heart. J Biol Chem 283:20045–20052.

Mahato RI, Cheng K, Guntaka RV (2005) Modulation of gene expression by antisense and antigene oligodeoxynucleotides and small interfering RNA. Expert Opin Drug Deliv 2:3–28.

Nabholtz JM, Gligorov J (2005) Docetaxel/trastuzumab combination therapy for the treatment of breast cancer. Expert Opin Pharmacother 6:1555–1564.

Ogihara T (2003) The combination therapy of hypertension to prevent cardiovascular events (COPE) trial: Rationale and design. Hypertens Res 28:331–338.

Pillai RS, Bhattacharyya SN, Filipowicz W (2007) Repression of protein synthesis by miRNAs: How many mechanisms? Trends Cell Biol 17:18–126.

Pushparaj PN, Aarthi JJ, Manikandan J, Kumar SD (2008) siRNA, miRNA, and shRNA: In vivo applications. J Dent Res 87:992–1003.

Sun D, Melegari M, Sridhar S, Rogler CE, Zhu L (2006) Multi-miRNA hairpin method that improves gene knockdown efficiency and provides linked multi-gene knockdown. Biotechniques 41:59–63.

Wang Z, Luo X, Lu Y, Yang B (2008) miRNAs at the heart of the matter. J Mol Med 86:771–783.

Xia H, Mao Q, Paulson HL, Davidson BL (2002) siRNA-mediated gene silencing in vitro and in vivo. Nat Biotechnol 20:1006–1010.

Xia XG, Zhou H, Xu Z (2006) Multiple shRNAs expressed by an inducible pol II promoter can knock down the expression of multiple target genes. Biotechniques 41:64–68.

Xiao J, Luo X, Lin H, Xu C, Gao H, Wang H, Yang B, Wang Z (2007a) MicroRNA miR-133 represses HERG K^{+}channel expression contributing to QT prolongation in diabetic hearts. J Biol Chem 282:12363–12367.

Xiao J, Yang B, Lin H, Lu Y, Luo X, Wang Z (2007b) Novel approaches for gene-specific interference via manipulating actions of microRNAs: examination on the pacemaker channel genes HCN2 and HCN4. J Cell Physiol 212:285–292.

Yang B, Lin H, Xiao J, Luo X, Li B, Lu Y, Wang H, Wang Z (2007) The muscle-specific microRNA miR-1 causes cardiac arrhythmias by targeting GJA1 and KCNJ2 genes. Nat Med 13:486–491.

Yang B, Lu Y, Wang Z (2008) Control of cardiac excitability by microRNAs. Cardiovasc Res 79:571–580.

Zhang Y, Lu Y, Wang N, Lin H, Pan Z, Gao X, Zhang F, Zhang Y, Xiao J, Shan H, Luo X, Chen G, Qiao G, Wang Z, Yang B (2009) Control of experimental atrial fibrillation by microRNA-328. Science (in revision)

Chapter 3
Synthetic Canonical miRNA Technology

Abstract Synthetic Canonical miRNAs are exogenously applied miRNAs in the form of mature double-stranded constructs or precursor hairpin constructs that are identical in sequence to their counterpart endogenous miRNAs. For convenience, I use the abbreviation SC-miRNAs throughout for Synthetic Canonical miRNAs. SC-miRNAs, once introduced into cells, produce seemingly identical gene regulation and cellular functions as do their endogenous miRNA counterparts. The SC-miRNA approach belongs to the "miRNA-targeting" and "miRNA-gain-of-function" strategy. This strategy conforms to the concept of 'miRNA as a Regulator of a Cellular Function'.

3.1 Introduction

miRNA-gain-of-function is an essential way for gain-of-knowledge about miRNA targets and functions. There is an array of approaches for achieving miRNA-gain-of-function: exogenously transfected miRNAs, miRNA transgenic mice and miRNA mimics. Among them, exogenously transfected miRNAs that can mimic the actions of endogenous canonical miRNAs is the most fundamental and commonly used miRNA-gain-of-function approach. Canonical miRNAs are defined as non-coding RNAs that meet the following criteria.

1. Mature miRNA should be expressed as a distinct transcript of ~22 nts that is detectable by RNA (Northern) blot analysis or other experimental means such as cloning from size-fractionated small RNA libraries.
2. Mature miRNA should originate from a precursor with a characteristic secondary structure, such as a hairpin or fold-back, which does not contain large internal loops or bulges. Mature miRNA should occupy the stem part of the hairpin.
3. Mature miRNA should be processed by Dicer, as determined by an increase in accumulation of the precursor in Dicer-deficient mutants.
4. In addition, an optional but commonly used criterion is that mature miRNA sequence and predicted hairpin structure should be conserved in different species.

Z. Wang, *MicroRNA Interference Technologies*,
DOI: 10.1007/978-3-642-00489-6_3,

Some people call SC-miRNAs "miRNA mimics" but this name is a bit confusing, as it sounds like the synthetic canonical miRNAs but not real miRNAs although they are synthesized exactly like endogenous miRNAs. This is as if we PCR synthesize p53 cDNA; we call it a p53 gene but not a p53 mimic. Further, there is another miRNAi technology called "miRNA mimics" (introduced in Chap. 4): (Xiao et al. 2007b; Wang et al. 2008). To avoid confusion, it is more logical to use "SC-miRNAs" to define this type of miRNAs.

3.2 Protocols

There are two ways to apply the SC-miRNA approach. One is through the introduction of a double-stranded miRNA mimetic or a synthetic canonical miRNA (SC-miRNA), equivalent to the endogenous Dicer product and analogous in structure to an siRNA (Soutschek et al. 2004). An alternative strategy for therapeutic miRNA replacement is a gene therapy approach. This involves expression of a short hairpin RNA (shRNA) from a polymerase II or III promoter, in a DNA or viral vector, which is then processed by Dicer before loading into RISC. Advantages to this approach are the potential for more persistent silencing compared to double-stranded miRNA mimetics, as well as the ease of expressing multiple miRNAs from one transcript. A step-by-step protocol for designing SC-miRNAs and validating SC-miRNAs is detailed below.

3.2.1 Designing SC-miRNAs

3.2.1.1 Double-Stranded SC-miRNA

1. Select a miRNA of interest for your study;
2. Obtain the sequences of that miRNA from miRNA database, such as miRBase (http://microrna.sanger.ac.uk/sequences/search.shtml). You can use either mature or precursor miRNA sequence. Double-stranded miRNA molecules mimic the Dicer cleavage product and hairpin precursor miRNAs can be processed by Dicer once delivered into cells. Do not forget to collect the oligonucleotide sequences for both guide and passenger strands;
3. Synthesize guide strand and passenger strand separately, either by yourself if you are equipped with a DNA/RNA synthesizer or by commercial services provided by several companies around the world;
4. Anneal guide and passenger strands to form a duplex SC-miRNA. First mix the two synthetic oligonucleotides at an equal molar concentration in an Eppendorf tube, then incubate the sample in a heat block at 95°C for 5 min and gradually cool the sample to room temperature (22–23°C);
5. Store the SC-miRNA construct at −80°C for future use;

6. Construct a negative control SC-miRNA for verifying the effects and specificity of the effects of the SC-miRNA. This negative control SC-miRNA should be designed based on the sequence of the SC-miRNA; simply modifying the SC-miRNA to contain ~5-nts mismatches at the 5′ portion "seed site". Such modified or mutated SC-miRNA would lose the ability to bind the target mRNA with sufficient affinity and is thus rendered incapable of eliciting the action of a miRNA. A negative control SC-miRNA in duplex form is formed through the annealing procedures described above;
7. Double-stranded SC-miRNA may be chemically modified to improve nuclease stability.

Many 2′-sugar modifications have been shown to confer this property, including 2′-O-Me, 2′-F, 2′-deoxy, 2′-MOE and LNA, which have been evaluated and are tolerated to varying degrees in either the sense or antisense strands (Chiu and Rana 2003; Czauderna et al. 2003; Prakash et al. 2005; Kraynack and Baker 2006; Allerson et al. 2005; Layzer et al. 2004; Harborth et al. 2003; Braasch et al. 2003; Morrissey et al. 2005a; Amarzguioui et al. 2003). Introduction of phosphorothioate internucleotidic linkages, which can promote plasma protein binding and delay renal clearance of single stranded oligonucleotides, in addition to conferring nuclease stability (Altmann et al. 1996; Hoke et al. 1991), is tolerated by siRNAs in cell culture (Chiu and Rana 2003; Czauderna et al. 2003; Braasch et al. 2003; Amarzguioui et al. 2003) and a beneficial effect on in vivo delivery has been reported (Braasch et al. 2004). The first demonstration of in vivo systemic delivery to the liver was achieved through high pressure intravenous injection, or hydrodynamic delivery of unmodified siRNAs (McCaffrey et al. 2002). Alternative strategies to promote cellular uptake and prevent renal clearance after systemic delivery have involved conjugation of the siRNA to moieties such as cholesterol (McCaffrey et al. 2002), or formulation in liposomes (Zimmermann et al. 2006; Morrissey et al. 2005b; Ge et al. 2004), although uptake and activity have only been demonstrated in the liver and lungs.

3.2.1.2 SC-miRNA Expression Vectors

Whereas administration of miRNA mimics is expected to provide transient recovery of under-expressed miRNAs, it is possible to express the mature miRNA form by using short hairpin RNAs (shRNAs) driven by Pol III promoters (Brummelkamp et al. 2002a; Paddison et al. 2002). When delivered within the context of an integrating vector system, Pol III-driven miRNA hairpins can be stably expressed. The Pol III-driven shRNA system is attractive since shRNAs have a simple design resembling pre-miRNAs and many Pol III promoters provide abundant expression from well-defined transcription start and termination sites. Although abundant expression ensures effective target knockdown, it can also saturate the exportin-5 pathway of endogenous miRNAs with fatal consequences (Saetrom et al. 2006). Therefore, the more prudent approach for stable expression of miRNAs is to opt for induced or temporal expression (Snove and Rossi 2006).

To ensure the effective target knockdown, the miRNA duplex should be released from the Pol III-driven shRNAs by the action of Dicer. One approach to effective shRNA processing is to use hairpins with longer (25–29 nts) stems (Kim et al. 2005; Siolas et al. 2005). A different approach for designing therapeutic miRNA expression vectors, which circumvents the use of artificial short and long hairpin, is to clone the entire natural pre-miRNA hairpin into a Pol III expression vector (Boden et al. 2004; Chung et al. 2006; Zeng et al. 2002, 2005; Zhou et al. 2005). The basic premise underlying this approach is that endogenous miRNAs have evolved to be good substrates for effective processing by miRNA biogenesis enzymes.

In addition to RNA Pol III systems, which express miRNAs within the context of a shRNA, other miRNA expression systems have been reported that are modeled on the processing of primary miRNA transcripts (Paddison et al. 2002; Boden et al. 2004; Zeng et al. 2002). Overexpression of endogenous miRNAs can be achieved via expression systems that use viral or liposomal delivery (Voorhoeve et al. 2006; Chung et al. 2006; Stegmeier et al. 2005). An efficient approach might be to express the miRNA or siRNA as hairpins using expression vectors containing polymerase III promoters such as H1 (Brummelkamp et al. 2002b) or U6 (Miyagishi and Taira 2002) promoters. miRNAs can also be expressed from vectors containing polymerase II promoters, which is CMV controlled expression vectors (Chung et al. 2006; Stegmeier et al. 2005; Xia et al. 2002). In this latter case, it was described that miRNA-mediated silencing was more efficient when expressing the miRNA in a pri-miRNA form, thus including miRNA flanking sequences and the hairpin structure, as shown for miR-30 (Stegmeier et al. 2005) and miR-155 (Chung et al. 2006).

Modified adenovirus or adeno-associated virus (AAV) vectors have been effective for gene delivery to the liver (Uprichard et al. 2005), brain (Xia et al. 2002) and heart (Zhang et al. 2008). Figure 3.1 illustrates the protocols for generating adenovirus expression vector expressing miR-328 for studying the arrhythmogenic potential of this miRNA in hearts of dogs and rats (Zhang et al. 2008). They are limited, however, by the immune response to the adenovirus and limited in the tissues that they can efficiently infect. Lentiviruses have also been investigated as vectors for shRNA expression, which have the potential to produce even more stable, long-lasting silencing as they integrate into the genome (Buchschacher and Wong-Staal 2000). ES cells infected with shRNA-expressing lentivirus have been used to create knockdown mice (Rubinson et al. 2003). However, the potential for insertional mutagenesis may be a concern for therapeutic use of lentiviral vectors.

For detailed step-by-step protocol, please refer to Chap. 5.

3.2.2 Validating SC-miRNAs

Reporter genes, such as the luciferase reporter gene, have been applied to confirm predicted miRNA binding sites making them a useful tool for the initial analysis of

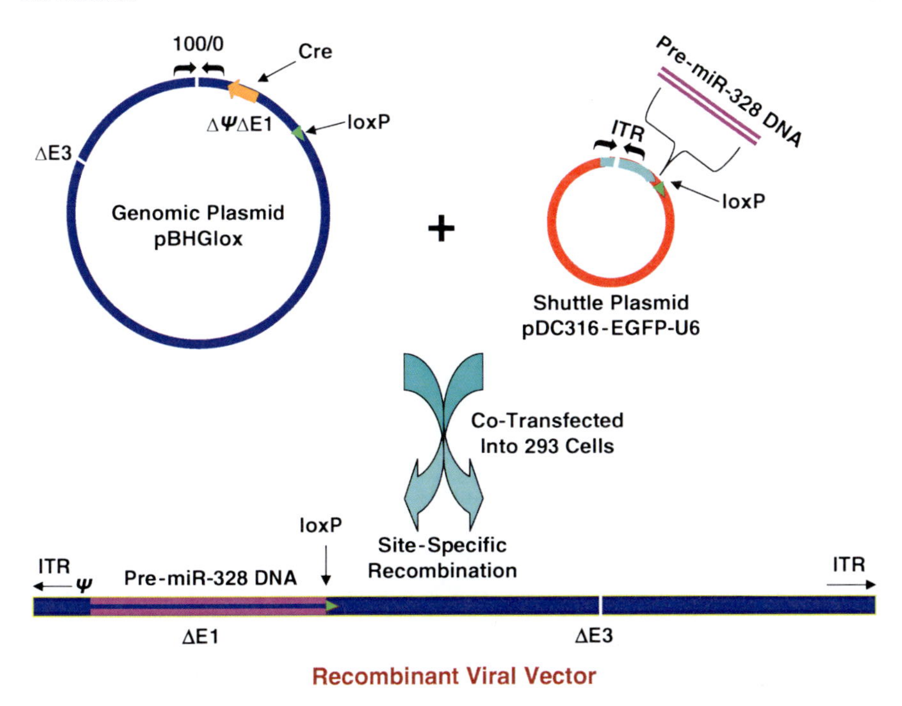

Fig. 3.1 Schematic illustration of construction of adenovirus vector carrying pre-miR-328. Rat miR-328 precursor DNA (5′-GGATCCgACCCCGTCCCCCCGTCCTCCCCGAGTCCCTCTT-TCGTAGATGTCGGGGACCGGGAGAGACGGGAAGGCAGGGGACAGGGGTTTAttttttAAG-CTT-3′) is inserted into adenovirus shuttle plasmid pDC316-EGFP-U6. pDC316-EGFP-U6 is then cotransfected with the infectious adenovirus genomic plasmid pBHGloxΔE1,3Cre into 293 cells by lipofectamine. Following cotransfection of these two DNAs, homologous recombination will occur to generate a recombinant adenovirus in which pre-miR-328 is incorporated into the viral genome, replacing the Δ*E*1 region

potential miRNA-binding sites. However, without evidence to demonstrate that the endogenous miRNA is changing the expression of the endogenous gene, it is difficult to know whether these results are biologically relevant. This will require that the antibody against the predicted products is available for Western blot analysis. Since predicted miRNA targets are a good roadmap for possible targets, this is an effective approach to take to verify activity of the miRNA inhibitors. Often, quantification of targeted mRNA is also required for better understanding of the actions of a SC-miRNA.

3.2.2.1 Luciferase Reporter Assay

1. Insert the 3′UTR containing the binding site of a SC-miRNA into the multiple cloning sites downstream the luciferase gene in a luciferase vector, such as

the pMIR-REPORT luciferase miRNA expression reporter vector provided by Ambion;
2. Transfect the luciferase vectors (1 μg firefly luciferase vector and 0.1 μg TK-driven Renilla luciferase expression vector), along with the SC-miRNA or negative control SC-miRNA, into the cells (at a density of 1×10^5/well) selected for testing. Transfection can be done with lipofectamine 2000 or other lipid carriers. For all experiments, transfection should be done 24 h after starvation of cells in serum-free medium. Alternatively and ideally, miR-Mimic can be constructed into virus vectors (such as adenovirus) to enhance the efficiency of delivery into the cells;
3. Measure luciferase activities with the dual luciferase reporter assay kit (Promega) on a luminometer, 36–48 h after transfection.

3.2.2.2 Verification of Downregulation of Cognate Target Protein

SC-miRNAs, like natural and endogenous miRNAs, silence target genes post-transcriptionally when introduced into cells. Downregulation of target genes at the protein level is a characteristic of SC-miRNA actions. But this needs to be confirmed for each SC-miRNA under study. Western blot analysis is a commonly used method for verification of downregulation of cognate target protein, though other techniques like immunostaining are also convenient for the purpose (Yang et al. 2007; Luo et al. 2007, 2008; Xiao et al. 2007a, b). Western blotting procedures are those routines you can find in any laboratory involved with biomedical research. Do not forget to distinguish among the membrane protein, cytosolic protein and nuclear protein samples for your particular need. In addition to Western blot and immunostaining, functional assays should also be employed when needed, such as enzyme activity assay (Xu et al. 2007), cell growth and death (Xu et al. 2007) and patch-clamp recordings for ion channels as target genes (Yang et al. 2007; Luo et al. 2007, 2008; Xiao et al. 2007a, b).

3.2.2.3 Quantification of mRNA and miRNA

The purposes of RNA quantification described in this section are twofold: one for verifying overexpression of the SC-miRNA introduced into cells and the other for measuring the possible alterations of the cognate target mRNA produced by the SC-miRNA. The traditional method of validation has been Northern blotting analysis. Northern blotting is an effective method to visualize both the precursor and the mature miRNA. The sensitivity of Northern blotting, however, is not sufficiently high for detecting some low-abundance miRNAs. As an alternative to Northern blotting, real-time quantitative reverse transcriptase-polymerase chain reaction (qRT-PCR) has been extremely effective in these validation studies. The methods can be used for quantifying both the precursor and mature miRNAs. Another type of validation is to visualize the expression of the miRNA to locate the cellular and subcellular distribution using in situ hybridization.

Northern Blotting Analysis

The following protocols for Northern blot can be found in Protocol Online – your Lab's Reference Book (Contributed by Zuyuan Qian) [http://www.protocol-online.org/prot/Molecular_Biology/RNA/Northern_Blotting/].

1. Agarose/Formaldehyde Gel Electrophoresis
   (a) Prepare gel: Dissolve 0.75 g agarose in 36 ml water and cool to 60°C in a water bath. When the flask has cooled to 60°C, place in a fume hood and add 5 ml of 10 × E running buffer and 9 ml formaldehyde. Pour the gel and allow it to set. Remove the comb, place the gel in the gel tank and add sufficient 1 × E running buffer to cover to a depth of ∼1 mm.
   (b) Prepare sample: Adjust the volume of each RNA sample to 6 μl with water, then add 2.5 × 6 μl freshly prepared sample denaturation mix. Mix by vortexing, microcentrifuge briefly to collect liquid and incubate 15 min at 55°C. Cool on ice for 2 min, then add 2 μl loading dye mix.
   (c) Run gel: Run the gel in 1 × E running buffer at 100 V for 10 min, then at 65 v for 90 min
2. Transfer of RNA from Gel to Membrane
   (a) Prepare gel for transfer: Place the gel in an RNase-free dish and rinse with sufficient changes of deionized water to cover the gel for 4 × 20 min.
   (b) Transfer RNA from gel to membrane:
      – Fill the glass dish with enough 20 × SSPE.
      – Cut 2 pieces of Whatman 1MM paper, place it on the glass plate and wet it with 20 × SSPE.
      – Place the gel on the filter paper and squeeze out air bubbles by rolling with a glass pipet.
      – Cut four strips of plastic wrap and place over the edges of the gel.
      – Cut a piece of nylon membrane (MSI, Catalog #N00HY320F5) just large enough to cover the gel and wetted in water. Place the wetted membrane on the surface of the gel. Try to avoid getting air bubbles under the membrane.
      – Flood the surface of the membrane with 20 × SSPE. Cut 5 sheets of whatman 3MM paper to the same size as the membrane and place on top of the membrane.
      – Place paper towels on top of the whatman 3MM paper to a height of ∼6cm and add a weight to hold everything in place.
      – Leave overnight.
   (c) Prepare membrane for hybridization: Remove paper towels and filter papers and recover the membrane and flattened gel. Mark in pencil the position of

the wells on the membrane and ensure that the up-down and back-front orientation are recognizable. Rinse the membrane in 5 × SSPE, then place it on a sheet of Whatman 3MM paper and allow to dry. Place RNA-side-down on a UV transilluminator (254 nm wavelength) and irradiate for an appropriate length of time.

3. Hybridization Analysis

   (a) Prepare DNA or RNA probe (>108 dpm μg^{-1}):
   The probe labeled with Ridiprimer DNA labelling system (Amersham LIFE SCIENCE):
      - Dilute the DNA to be labeled to a concentration of 2.5–25 ng in 45 μl of sterile water.
      - Denature the DNA sample by heating to 95–100°C for 5 min in a boiling water bath.
      - Centrifuge briefly to bring the contents to the bottom of the tube and put on ice for 10 min.
      - Add the denatured DNA to the labeling mix and reconstitute the mix by gently flicking the tube until the blue color is evenly distributed.
      - Add 5 μl of Redivue [^{32}P] dCTP and mix by gently pipetting up and down. Centrifuge briefly to bring the contents to the bottom of the tube, then incubate at 37°C for 30 min.
      - The probe is purified using ProbeQuantTMG-50 micro columns (Amersham Pharmacia Biotech):
      - G-50 micro column preparation. Resuspend the resin in the column by vortexing, loosen the cap one-fourth turn and snap off the bottom closure. Place the column in a 1.5 ml screw-cap microcentrifuge tube for support, then pre-spin the column for 1 min at 3,000 rpm in an Eppendorf model 5415C.
      - Place the column in a new 1.5 ml tube and slowly apply 50 μl of the sample to the top-center of the resin, being carful not to disturb the resin bed. Spin the column at 3,000 for 2 min. The purified sample is collected in the bottom of the support tube.

   (b) Hybridization:
      - Pre-hybridization: Wet the membrane in the 5 × SSPE and place it RNA-side-up in a hybridization tube and add 5 ml pre-hybridization solution, then place the tube in the hybridization oven and incubate with rotation 6 hr at 42°C for DNA probe or 60°C for RNA probe.
      - Hybridization: Double-stranded probe was denatured by heating in a water bath for 10 min at 100°C, then transferred to ice. Pipet the desired volume of probe into the hybridization tube and continue to incubate with rotation overnight at 42°C for DNA probe or 60°C for RNA probe.

   (c) Autoradiography:
      - The membrane was washed twice for 5–10 min with wash-buffer at room temperature and twice for 15 min at 65°C with prewarmed (65°C) wash-buffer.

- Remove final wash solution and rinse membrane in 5 × SSPE at room temperature. Blot excess liquid and cover in UV-transparent plastic wrap. Do not allow membrane to dry out if it is to be reprobed.
- Blot was exposed at −80°C unsing Kodak XAR film and x-ray intensifying screens.

Real-Time qRT-PCR

The first real-time PCR method to quantify miRNA precursors was developed by Schmittgen and colleagues (Schmittgen et al. 2004). Shortly thereafter, Chen and colleagues (Chen et al. 2005) developed a real-time PCR assay to quantify mature miRNA. Mature miRNA is the active species and exerts its activity by binding to the 3′UTR of mRNA. Quantification of the active, mature miRNA, rather than the inactive, pre-miRNA, is generally preferred. The methods described below are mainly based on those reported by Schmittgen et al. (2004, 2008) and Chen et al. (2005). TaqMan-based real-time quantification of miRNAs includes two steps, stem–loop RT and real time PCR. Stem–loop RT primers bind to the 3′ portion of miRNA molecules and are reverse transcribed with reverse transcriptase. Then, the RT product is quantified using conventional TaqMan PCR that includes miRNA-specific forward primer, reverse primer and dye-labeled TaqMan probes. You can either simply use the TaqMan PCR kit for miRNA provided by Applied Biosystems Inc or run the PCR using your own primers and reagents. If you chose the latter, you may want to follow the brief instructions below. I encourage readers to consult the original research articles for more detailed information about the procedures.

1. Primer design. Obtain pre-miRNA sequences from the miRNA registry [http://microrna.sanger.ac.uk/registry/; the Wellcome Trust Sanger Institute]. Precursor sequences listed in the miRNA registry do not represent the identical pre-miRNA sequence and contain additional nts that are 5′ and 3′ of the hairpin. Please follow the instructions presented in the article by Schmittgen et al. (2008) to acquire more precise pre-miRNA sequences. For pre-miRNAs, both forward and reverse primers must be located within the hairpin sequence of the pre-miRNAs. For mature miRNAs, a stem-loop RT primer must be used. Primer Express version 2.0 (Applied Biosystems, Foster City, CA) is convenient for the task;
2. TaqMan probe design. The probes should be designed to anneal to the loop portion of pre-miRNAs and are typically designed to have a T_m that is 10°C higher than the primers;
3. Reverse transcription;
4. Real-time PCR. relative quantification, absolute quantification.

In Situ Hybridization (ISH)

The protocols described below are essential from those reported by Thompson et al. (2007).

1. Probe design and controls. In most cases, synthetic RNA oligonucleotide probes with 20 nt of complementarity to a specific target miRNA can be used and the conditions presented here are optimized for this probe size. However, probes with 21 or 22 nt complementary to a target miRNA appear to function similarly (Deo et al. 2006). The fluorescein-labeled RNA probes include two additional nucleotides at the 5′ end that do not match the target miRNA. These bases are not required for probe function but they are included to potentially improve access of the anti-fluorescein antibody to the 5′ end fluorescein. Since the TMAC-based wash conditions are not sensitive to sequence composition, the choice of which 20 nt sequence within a miRNA longer than 20 nt should be used for probe generation appears to be arbitrary. Since many miRNAs are part of gene families that are composed of miRNAs that differ at only one or a few positions, all probe designs should be compared against miRBase (http://microrna.sanger.ac.uk/) (Griffiths-Jones 2006) to identify potential unanticipated miRNA matches. In general, if it is desirable to distinguish between closely related miRNAs, probes should be positioned on the target miRNAs so as to maximize the number of mismatches with non-target miRNAs and so that potential duplexes between probe and non-target miRNAs will be as short as possible (to reduce cross-hybridization). However, the short length of miRNAs limits the opportunities for probe specificity optimization. In addition, it is unlikely that any miRNA ISH procedure will be able to distinguish between miRNAs that differ by only a single nucleotide at one end (e.g., miR-128a and miR-128b).

To confirm the sequence specificity of ISH, control probes can include mutations that create mismatches with the target miRNA at 1 to 3 internal positions (Wienholds et al. 2005; Deo et al. 2006). It is important to avoid mismatch mutations that would allow a probe to hybridize to other members of a miRNA gene family unintentionally.

To detect probe trapping or other types of sequence-independent probe binding that can lead to elevated background, a reversed or scrambled sequence RNA probe can be used (after comparison to miRBase to check for unanticipated targets). One limitation of reversed or scrambled sequence controls is that it is not easy to verify that they are functional probes, since they have no endogenous target.

An alternative is to use a probe that detects a known miRNA that is not expressed in the tissue of interest (or which is expressed in a distinct pattern within the tissue of interest). In this case, the integrity and function of the control probe can be confirmed by ISH using the appropriate tissue. When comparing ISH using control probes to ISH with probes for miRNAs of interest, it is important to use identical processing for all probes, including the duration of AP staining or autoradiography. In all cases, comparing and contrasting a probe-specific hybridization signal with the non-specific hybridization signal from one or more control probes provides the user with some level of confidence in the anatomical pattern of gene expression observed.

Beyond probe sequence design, anatomical replication of hybridization results within and across tissue sections/slides is also important. Multiple tissue sections are placed per microscope slide and if the anatomy is replicated in these tissue sections, then replicate hybridization signals should be detected in each section/slide, suggesting that the hybridization results are representative of cells expressing the miRNA being evaluated. Similarly if there is symmetry of gene expression (e.g., right and left brain hemispheres), one would predict that hybridization results would display similar anatomical symmetry thereby demonstrating an anatomical consistency of spatial expression patterns.

It is also possible to compare the expression patterns of mature miRNAs determined by miRNA ISH with the expression patterns of the corresponding miRNA primary transcripts, as detected by conventional ISH methods (Deo et al. 2006). The expression of the mature miRNA and the primary transcript should overlap, although identical miRNAs can be encoded by multiple genes, potentially complicating ISH analysis of the primary transcripts. In addition, recent observations indicate that miRNA maturation or stability may be regulated independently of primary transcript synthesis (Wulczyn et al. 2007; Obernosterer et al. 2006; Thomson et al. 2006), so that the mature miRNA expression pattern may not match the primary transcript pattern in all cases.

2. RNA oligonucleotide probe preparation. RNA oligonucleotide probes can be custom synthesized commercially (Invitrogen). For nonradioactive ISH, RNA oligonucleotides are synthesized with a 5′ end fluorescein modification. In most cases, gel purification of fluorescein probes as described below is essential to reduce background on tissue sections (apparently from unincorporated fluorescein-labeled nucleotide). For radioactively-labeled ISH, RNA oligonucleotides are custom synthesized without any 5′ end modification (5′ hydroxyl) and are enzymatically labeled at the 5′ end with ^{33}P using T4 DNA Kinase, as described below. Note that 5′ fluorescein-modified RNA oligonucleotides cannot be enzymatically labeled with T4 DNA Kinase.
3. Gel purification of fluorescein-labeled RNA oligonucleotides. Load 20 μg of fluorescein-labeled RNA oligonucleotide onto an 18% polyacrylamide/TBE gel. After electrophoresis, identify the RNA oligonucleotide band by fluorescence from the incorporated fluorescein (using a standard UV transilluminator). Isolate the highest molecular weight band. Cut out RNA in smallest band possible and transfer to 100 μl of diethyl pyrocarbonate (DEPC)-treated 1X phosphate buffered saline (PBS; see section 3 for descriptions of solutions). Crush gel into as small pieces as possible and transfer to a microfuge tube. Incubate overnight at 37°C. Centrifuge at ∼11,000 g for 15 min and collect supernatant. To precipitate probe: add 1/10 volume DEPC-treated 3 M NaOAc, 2.5 volumes 100% ethanol and store at −20°C for 1 h. Centrifuge at ∼11,000 g for 15 min. Remove ethanol and allow pellet to air dry. Pellet will be visibly yellow. Resuspend in 10 μl of RNase-free water and store at −20°C. Integrity of the purified probe can be confirmed by gel elecrophoresis using a small amount of the purified probe

on an 18% polyacrylamide/TBE gel (the unpurified probe can be used for size comparison).

4. 5′ end labeling of RNA oligonucleotides with ^{33}P. Incubate 5–10 pmol of RNA oligonucleotide and ~100 μCi of 1,000 Ci/mmol $^{33}P-\gamma ATP$ with T4 DNA Kinase according to the enzyme manufacture's recommendations (typically 30–45 min at 37°C). ^{33}P-labeled RNA oligonucleotides are purified from unincorporated $^{33}P-\gamma ATP$ using P6 BioRad Spin columns according to the manufacture's directions. Typical incorporation percentages are ~50%. Since incorporation rates are similar across different RNA oligonucleotides, we routinely use $\sim 0.5-1.0\times 10^6$ cpm of labeled RNA oligonucleotide per 40 μl of hybridization buffer.
5. Preparation of tissue sections for miRNA ISH. As with conventional mRNA ISH, tissue can be fixed at the time of collection by immersion or perfusion, or fresh-frozen tissue sections can be used. Embryos or small pieces of tissue can be fixed by overnight immersion in 4% paraformaldehyde (PFA) in PBS at 4°C. It is better to fix large adult tissues such as brain by perfusion with PFA. After fixation, rinse tissues in 1X PBS, dehydrated in 15% sucrose until the tissue sinks and embedded in OCT (Fisher). Cut tissue sections (12 μm) using a cryostat and transfer to SuperFrost/plus slides (Fisher). Store slides at −20°C until ISH. Alternately, for fresh-frozen tissue, collect tissues as quickly as possible and immediately freeze them in dry ice-chilled isopentane baths (−35°C). Store the frozen tissues at −80°C until sectioning. Frozen tissue sections (10–14 μm thick) are thaw mounted onto clean, SuperFrost microscope slides and stored at −80°C until ISH.
6. Preparation of Cultured Cells for miRNA ISH. Place cells on poly-L-Lysine and laminin-coated microscope slides. This coating helps retain cells during the ISH processing steps. Coat microscope slides with poly-L-Lysine ($10\,\mu g\,ml^{-1}$) in 1X PBS for 3.5 h, followed by air drying in a laminar flow hood. Then rinse slides three times with distilled water and store them at 4°C (for up to two weeks). Prior to use, coat slides with mouse Laminin ($2\,\mu g\,ml^{-1}$, Invitrogen) overnight, wash with 1X PBS twice prior to plating cultured cells onto the slides in appropriate media. At the desired time, wash cells with 1X D-PBS (Invitrogen) for 5 min and fix in PFA for 20 min. Rinse the fixed cells three times with DEPC-treated 1X PBS and store in DEPC-treated 1X PBS at 4°C until in situ hybridization (cells can be stored for a few days).

3.3 Principle of Actions

The mechanism for the action of SC-miRNAs is straightforward; SC-miRNAs simply act as endogenous miRNAs through interacting with Argonaute proteins in miRISC and guiding miRISC to the target genes via partial base-pairing between the guide strand of SC-miRNA and the cognate target mRNA (mainly the 3′UTR of mRNAs). SC-miRNAs can then work by two modes-mRNA cleavage or translational repression without RNA cleavage. A single SC-miRNA has the potential to recognize several target sequences in the 3'UTR and cause translation inhibition of many different genes.

3.4 Applications

1. To achieve "miRNA-gain-of-function". Perturbation of miRNA expression, both overexpression and silencing, is a powerful approach to study miRNA functions and to validate miRNA targets. Transient overexpression of miRNAs in cell-based assays can be achieved by transfection and in tissues by in vivo gene transfer techniques. This "miRNA-gain-of-function" strategy has become indispensable and most essential in fundamental research of miRNAs and of miRNA-related biological processes. SC-miRNA is a direct and straightforward approach to achieve miRNA-gain-of-function. Indeed, ever since the discovery of the first miRNA in 1993 (Lee et al. 1993), SC-miRNAs have been utilized in nearly all experiments involving miRNAs.
2. To achieve "miRNA replacement therapy". Some miRNAs are decreased in their expression levels in certain diseased tissues. Correction of deregulated expression of these endogenous miRNAs by 'replacement therapy' may be an efficient approach for management of the disorders. Introduction of synthetic miRNAs into cells as a "gain-of-function" approach has proven feasible under many conditions. There has already been a suggestion that replacement of miRNAs may be therapeutically beneficial in some cancers and other therapeutic opportunities will certainly be discovered in the future.

One way to achieve this is through the introduction of SC-miRNAs. For instance, miR-34a has recently attracted tremendous attention owing to its ability to mediate p53-induced apoptosis (Chang et al. 2007; Corney et al. 2007; He et al. 2007; Rokhlin et al. 2008) and its expresson is downregulated in cancer cells (Guglielmelli et al. 2007; Lodygin et al. 2008; Izzotti et al. 2008; Tryndyak et al. 2008). Data have now been available indicating that miR-34a directly targets the E2F3 mRNA and significantly reduces the levels of E2F3 protein in neuroblastoma (Welch et al. 2007; Tryndyak et al. 2008). Similar results were obtained by an independent group in two colon cancer cell lines, HCT 116 and RKO. The authors showed that introduction of miR-34a into the cells elicits a profound inhibition of cell proliferation accompanying the downregulation of the E2F family (Tazawa et al. 2007). In addition, the same group also found that miR-34a causes upregulation of the HBP1 gene, which is associated with oncogenic RAS-induced premature senescence. A recent study further demonstrated that ectopic expression of miR-34a reduces the expression of cyclin D1 (CCND1) and cyclin-dependent kinase 6 (CDK6) at both mRNA and protein levels (Sun et al. 2008).

3.5 Advantages and Limitations

The SC-miRNA approach has the advantages of being simple, straightforward and cost-effective.

This approach for therapeutic miRNA replacement will face all the same hurdles that siRNA therapeutics currently face, primarily the problem of systemic delivery to tissues.

References

Allerson CR, Sioufi N, Jarres R, Prakash TP, Naik N, Berdeja A, Wanders L, Griffey RH, Swayze EE, Bhat B (2005) Fully 2′-modified oligonucleotide duplexes with improved in vitro potency and stability compared to unmodified small interfering RNA. J Med Chem 48:901–904.

Altmann KH, Dean NM, Fabbro D, Freier SM, Geiger T, Haener R, Huesken D, Martin P, Monia BP, Muller M, Natt F, Nicklin P, Phillips J, Pieles U, Sasmor H, Moser H (1996) Second generation of antisense oligonucleotides. From nuclease resistance to biological efficacy in animals. Chimia 50:168–176.

Amarzguioui M, Holen T, Babaie E, Prydz H (2003) Tolerance for mutations and chemical modifications in a siRNA. Nucleic Acids Res 31:589–595.

Boden D, Pusch O, Silbermann R, Lee F, Tucker L, Ramratnam B (2004) Enhanced gene silencing of HIV-1 specific siRNA using microRNA designed hairpins. Nucleic Acids Res 32:1154–1158.

Braasch DA, Jensen S, Liu Y, Kaur K, Arar K, White MA, Corey DR (2003) RNA interference in mammalian cells by chemically-modified RNA. Biochemistry 42:7967–7975.

Braasch DA, Paroo Z, Constantinescu A, Ren G, Oz OK, Mason RP, Corey DR (2004) Biodistribution of phosphodiester and phosphorothioate siRNA. Bioorg Med Chem Lett 14:1139–1143.

Brummelkamp TR, Bernards R, Agami R (2002a). Stable suppression of tumorigenicity by virus-mediated RNA interference. Cancer Cell 2:243–247.

Brummelkamp TR, Bernards R, Agami R (2002b) A system for stable expression of short interfering RNAs in mammalian cells. Science 296:550–553.

Buchschacher Jr G.L, Wong-Staal F (2000) Development of lentiviral vectors for gene therapy for human diseases. Blood 95:2499–2504.

Chen C, Ridzon DA, Broomer AJ, Zhou Z, Lee DH, Nguyen JT, Barbisin M, Xu NL, Mahuvakar VR, Andersen MR, Lao KQ, Livak KJ, Guegler KJ (2005) Real-time quantification of microRNAs by stem-loop RT-PCR. Nucleic Acids Res 33:e179.

Chiu YL, Rana TM (2003) siRNA function in RNAi: A chemical modification analysis. RNA 9:1034–1048.

Chang TC, Wentzel EA, Kent OA, Ramachandran K, Mullendore M, Lee KH, Feldmann G, Yamakuchi M, Ferlito M, Lowenstein CJ, Arking DE, Beer MA, Maitra A, Mendell JT (2007) Transactivation of miR-34a by p53 broadly influences gene expression and promotes apoptosis. Mol Cell 26:745–752.

Chung KH, Hart CC, Al-Bassam S, Avery A, Taylor J, Patel PD, Vojtek AB, Turner DL (2006) Polycistronic RNA polymerase II expression vectors for RNA interference based on BIC/miR-155. Nucleic Acids Res 34:e53.

Corney DC, Flesken-Nikitin A, Godwin AK, Wang W, Nikitin AY (2007) MicroRNA-34b and MicroRNA-34c are targets of p53 and cooperate in control of cell proliferation and adhesion-independent growth. Cancer Res 67:8433–8438.

Czauderna F, Fechtner M, Dames S, Aygun H, Klippel A, Pronk GJ, Giese K, Kaufmann J (2003) Structural variations and stabilising modifications of synthetic siRNAs in mammalian cells. Nucleic Acids Res 31:2705–2716.

Deo M, Yu JY, Chung KH, Tippens M, Turner DL (2006) Detection of mammalian microRNA expression by in situ hybridization with RNA oligonucleotides. Dev Dyn 235:2538–2548.

Ge Q, Filip L, Bai A, Nguyen T, Eisen HN, Chen J (2004) Inhibition of influenza virus production in virus-infected mice by RNA interference. Proc Natl Acad Sci USA 101:8676–8681.

Griffiths-Jones S (2006) miRBase: The microRNA sequence database. Methods Mol Biol 342: 29–138.

Guglielmelli P, Tozzi L, Pancrazzi A, Bogani C, Antonioli E, Ponziani V, Poli G, Zini R, Ferrari S, Manfredini R, Bosi A, Vannucchi AM; MPD Research Consortium (2007) MicroRNA expression profile in granulocytes from primary myelofibrosis patients. Exp Hematol 35:1708–1718.

Harborth J, Elbashir SM, Vandenburgh K, Manninga H, Scaringe SA, Weber K, Tuschl T (2003) Sequence, chemical, and structural variation of small interfering RNAs and short hairpin RNAs and the effect on mammalian gene silencing. Antisense Nucleic Acid Drug Dev 13:83–105.

He L, He X, Lim LP, de Stanchina E, Xuan Z, Liang Y, Xue W, Zender L, Magnus J, Ridzon D, Jackson AL, Linsley PS, Chen C, Lowe SW, Cleary MA, Hannon GJ (2007) A microRNA component of the p53 tumour suppressor network. Nature 447:1130–1134.

Hoke GD, Draper K, Freier SM, Gonzalez C, Driver VB, Zounes MC, Ecker DJ (1991) Effects of phosphorothioate capping on antisense oligonucleotide stability, hybridization and antiviral efficacy versus herpes simplex virus infection. Nucleic Acids Res 19:5743–5748.

Izzotti A, Calin GA, Arrigo P, Steele VE, Croce CM, De Flora S (2009) Downregulation of microRNA expression in the lungs of rats exposed to cigarette smoke. FASEB J 23:806–812.

Kim DH, Behlke MA, Rose SD, Chang MS, Choi S, Rossi JJ (2005) Synthetic dsRNA Dicer substrates enhance RNAi potency and efficacy. Nat Biotechnol 23:222–226.

Kraynack BA, Baker BF (2006) Small interfering RNAs containing full 2′-Omethylribonucleotide-modified sense strands display Argonaute2/eIF2C2-dependent activity. RNA 12:163–176.

Layzer JM, McCaffrey AP, Tanner AK, Huang Z, Kay MA, Sullenger BA (2004) In vivo activity of nuclease-resistant siRNAs. RNA 10:766–771.

Lodygin D, Tarasov V, Epanchintsev A, Berking C, Knyazeva T, Körner H, Knyazev P, Diebold J, Hermeking H (2008) Inactivation of miR-34a by aberrant CpG methylation in multiple types of cancer. Cell Cycle 7:2591–2600.

Lee RC, Feinbaum RL, Ambros V (1993) The *C. elegans* heterochronic gene lin-4 encodes small RNAs with antisense complementarity to lin-14. Cell 75:843–854.

Luo X, Lin H, Lu Y, Li B, Xiao J, Yang B, Wang Z (2007) Transcriptional activation by stimulating protein 1 and post-transcriptional repression by muscle-specific microRNAs of I_{Ks}-encoding genes and potential implications in regional heterogeneity of their expressions. J Cell Physiol 212:358–367.

Luo X, Lin H, Pan Z, Xiao J, Zhang Y, Lu Y, Yang B, Wang Z (2008) Overexpression of Sp1 and downregulation of miR-1/miR-133 activates re-expression of pacemaker channel genes HCN2 and HCN4 in hypertrophic heart. J Biol Chem 283:20045–20052.

McCaffrey AP, Meuse L, Pham TT, Conklin DS, Hannon GJ, Kay MA (2002) RNA interference in adult mice. Nature 418:38–39.

Miyagishi M, Taira K (2002) U6 promoter-driven siRNAs with four uridine 30 overhangs efficiently suppress targeted gene expression in mammalian cells. Nat Biotechnol 20:497–500.

Morrissey DV, Blanchard K, Shaw L, Jensen K, Lockridge JA, Dickinson B, McSwiggen JA, Vargeese C, Bowman K, Shaffer CS, Polisky BA, Zinnen S (2005a) Activity of stabilized short interfering RNA in a mouse model of hepatitis B virus replication. Hepatology 41:1349–1356.

Morrissey DV, Lockridge J.A, Shaw L, Blanchard K, Jensen K, Breen W, Hartsough K, Machemer L, Radka S, Jadhav V, Vaish N, Zinnen S, Vargeese C, Bowman K, Shaffer CS, Jeffs LB, Judge A, MacLachlan I, Polisky B (2005b) Potent and persistent in vivo anti-HBV activity of chemically modified siRNAs. Nat Biotechnol 23:1002–1007.

Obernosterer G, Leuschner PJ, Alenius M, Martinez J (2006) Post-transcriptional regulation of microRNA expression. RNA 12:1161–1167.

Paddison PJ, Caudy AA, Bernstein E, Hannon GJ, Conklin DS (2002) Short hairpin RNAs (shRNAs) induce sequence-specific silencing in mammalian cells. Genes Dev 16:948–958.

Prakash TP, Allerson CR, Dande P, Vickers TA, Sioufi N, Jarres R, Baker BF, Swayze EE, Griffey RH, Bhat B (2005) Positional effect of chemical modifications on short interference RNA activity in mammalian cells. J Med Chem 48:4247–4253.

Rokhlin OW, Scheinker VS, Taghiyev AF, Bumcrot D, Glover RA, Cohen MB (2008) MicroRNA-34 mediates AR-dependent p53-induced apoptosis in prostate cancer. Cancer Biol Ther 7:1288–1296.

Rubinson DA, Dillon CP, Kwiatkowski AV, Sievers C, Yang L, Kopinja J, Rooney DL, Ihrig M.M, McManus M.T, Gertler FB, Scott M.L, Van Parijs L (2003) A lentivirus-based system to functionally silence genes in primary mammalian cells, stem cells and transgenic mice by RNA interference. Nat Genet 33:401–406.

Saetrom P, Snove O, Nedland M, Grunfeld TB, Lin Y, Bass, MB, Canon JR (2006) Conserved microRNA characteristics in mammals. Oligonucleotides 16:115–144.

Schmittgen TD, Lee EJ, Jiang J, Sarkar A, Yang L, Elton TS, Chen C (2008) Real-time PCR quantification of precursor and mature microRNA. Methods 44:31–38.

Schmittgen TD, Jiang J, Liu Q, Yang L (2004) A high-throughput method to monitor the expression of microRNA precursors. Nucleic Acids Res 32:e43.

Siolas D, Lerner C, Burchard J, Ge W, Linsley PS, Paddison PJ, Hannon GJ, Cleary MA (2005) Synthetic shRNAs as potent RNAi triggers. Nat Biotechnol 23:227–231.

Snove, O Jr, Rossi JJ (2006) Expressing short hairpin RNAs in vivo. Nat Methods 3:689–695.

Soutschek J, Akinc A, Bramlage B, Charisse K, Constien R, Donoghue M, Elbashir S, Geick A, Hadwiger P, Harborth J, John M, Kesavan V, Lavine G, Pandey RK, Racie T, Rajeev KG, Rohl I, Toudjarska I, Wang G, Wuschko S, Bumcrot D, Koteliansky V, Limmer S, Manoharan M, Vornlocher HP (2004) Therapeutic silencing of an endogenous gene by systemic administration of modified siRNAs. Nature 432:173–178.

Stegmeier F, Hu G, Rickles RJ, Hannon GJ, Elledge SJ (2005) A lentiviral microRNA-based system for single-copy polymerase II-regulated RNA interference in mammalian cells. Proc Natl Acad Sci USA 102:13212–13217.

Sun F, Fu H, Liu Q, Tie Y, Zhu J, Xing R, Sun Z, Zheng X (2008) Downregulation of CCND1 and CDK6 by miR-34a induces cell cycle arrest. FEBS Lett 582:1564–1568.

Tazawa H, Tsuchiya N, Izumiya M, Nakagama H (2007) Tumor-suppressive miR-34a induces senescence-like growth arrest through modulation of the E2F pathway in human colon cancer cells. Proc Natl Acad Sci USA 104:15472–15477.

Thomson JM, Newman M, Parker JS, Morin-Kensicki EM, Wright T, Hammond SM (2006) Extensive post-transcriptional regulation of microRNAs and its implications for cancer. Genes Dev 20:2202–2207.

Thompson RC, Deo M, Turner DL (2007) Analysis of microRNA expression by in situ hybridization with RNA oligonucleotide probes. Methods 43:153–161.

Tryndyak VP, Ross SA, Beland FA, Pogribny IP (2008) Down-regulation of the microRNAs miR-34a, miR-127, and miR-200b in rat liver during hepatocarcinogenesis induced by a methyl-deficient diet. Mol Carcinog 2008 Oct 21. [Epub ahead of print]

Uprichard SL, Boyd B, Althage A, Chisari FV (2005) Clearance of hepatitis B virus from the liver of transgenic mice by short hairpin RNAs. Proc Natl Acad Sci USA 102:773–778.

Voorhoeve PM, le Sage C, Schrier M, Gillis AJ, Stoop H, Nagel R, Liu YP, van Duijse J, Drost J, Griekspoor A, Zlotorynski E, Yabuta N, De Vita G, Nojima H, Looijenga LH, Agami R (2006) A genetic screen implicates miRNA-372 and miRNA-373 as oncogenes in testicular germ cell tumors. Cell 124:1169–1181.

Wang Z, Luo X, Lu Y, Yang B (2008) miRNAs at the heart of the matter. J Mol Med 86:771–783.

Welch C, Chen Y, Stallings RL (2007) MicroRNA-34a functions as a potential tumor suppressor by inducing apoptosis in neuroblastoma cells. Oncogene 26:5017–5022.

Wienholds E, Kloosterman WP, Miska E, Alvarez-Saavedra E, Berezikov E, de Bruijn E, Horvitz HR, Kauppinen S, Plasterk RH (2005) MicroRNA expression in zebrafish embryonic development. Science 309:310–311.

Wulczyn FG, Smirnova L, Rybak A, Brandt C, Kwidzinski E, Ninnemann O, Strehle M, Seiler A, Schumacher S, Nitsch R (2007) Post-transcriptional regulation of the let-7 microRNA during neural cell specification. FASEB J 21:415–426.

Xia H, Mao Q, Paulson HL, Davidson BL (2002) siRNA-mediated gene silencing in vitro and in vivo. Nat Biotechnol 20:1006–1010.

Xiao J, Luo X, Lin H, Xu C, Gao H, Wang H, Yang B, Wang Z (2007a) MicroRNA miR-133 represses HERG K^+ channel expression contributing to QT prolongation in diabetic hearts. J Biol Chem 282:12363–12367.

Xiao J, Yang B, Lin H, Lu Y, Luo X, Wang Z (2007b) Novel approaches for gene-specific interference via manipulating actions of microRNAs: Examination on the pacemaker channel genes HCN2 and HCN4. J Cell Physiol 212:285–292.

Xu C, Lu Y, Lin H, Xiao J, Wang H, Luo X, Li B, Yang B, Wang Z (2007) The muscle-specific microRNAs miR-1 and miR-133 produce opposing effects on apoptosis via targeting HSP60/HSP70 and caspase-9 in cardiomyocytes. J Cell Sci 120:3045–3052.

Yang B, Lin H, Xiao J, Luo X, Li B, Lu Y, Wang H, Wang Z (2007) The muscle-specific microRNA miR-1 causes cardiac arrhythmias by targeting GJA1 and KCNJ2 genes. Nat Med 13:486–491.

Zeng Y, Cai X, Cullen BR (2005). Use of RNA polymerase II to transcribe artificial microRNAs. Methods Enzymol 392:371–380.

Zeng Y, Wagner EJ, Cullen BR (2002) Both natural and designed micro RNAs can inhibit the expression of cognate mRNAs when expressed in human cells. Mol Cell 9:1327–1333.

Zimmermann TS, Lee AC, Akinc A, Bramlage B, Bumcrot D, Fedoruk MN, Harborth J, Heyes JA, Jeffs LB, John M, Judge AD, Lam K, McClintock K, Nechev LV, Palmer LR, Racie T, Rohl I, Seiffert S, Shanmugam S, Sood V, Soutschek J, Toudjarska I, Wheat AJ, Yaworski E, Zedalis W, Koteliansky V, Manoharan M, Vornlocher HP, MacLachlan I (2006) RNAi-mediated gene silencing in non-human primates. Nature 441:111–114.

Zhang Y, Lu Y, Wang N, Lin H, Pan Z, Gao X, Zhang F, Zhang Y, Xiao J, Shan H, Luo X, Chen G, Qiao G, Wang Z, Yang B (2008) Control of experimental atrial fibrillation by microRNA-328. Nat Med (in review).

Zhou H, Xia XG, Xu Z (2005) An RNA polymerase II construct synthesizes short-hairpin RNA with a quantitative indicator and mediates highly efficient RNAi. Nucleic Acids Res 33:e62.

Chapter 4
miRNA Mimic Technology

Abstract The miRNA Mimic technology (miR-Mimic) is an innovative approach for gene silencing. This approach generates non-natural double-stranded miRNA-like RNA fragments. Such a RNA fragment is designed to have its 5′ end bearing a partially complementary motif to the selected sequence in the 3′UTR unique to the target gene. Once introduced into cells, this RNA fragment, mimicking an endogenous miRNA, can bind specifically to its target gene and produce post-transcriptional repression, more specifically translational inhibition, of the gene. Unlike endogenous miRNAs, miR-Mimics act in a gene-specific fashion. The miR-Mimic approach belongs to the "miRNA-targeting" and "miRNA-gain-of-function" strategy and is primarily used as an exogenous tool to study gene function by targeting mRNA through miRNA-like actions in mammalian cells. The technology was developed by my research group (Department of Medicine, Montreal Heart Institute, University of Montreal) in 2007 [Xiao J, Yang B, Lin H, Lu Y, Luo X, Wang Z, J Cell Physiol 212:285–292, 2007; Xiao J, Lin H, Luo X, Chen G, Wang Z, Mol Cell, 2008]

4.1 Introduction

A general and unique feature of the miRNA-target RNA interaction is its imperfect complementarity between the miRNA guide strand and its target mRNA. Hence, miRNA guide strands usually form bulge structures due to mismatches with its target sequence. The sequence specificity for target recognition by the guide miRNA strand is determined by nucleotides 2–8 of its 5′ region, referred to as the "seed site" (Doench and Sharp 2004; Lewis et al. 2003). This short seed site, required for miRISC function, raises the potential for a single miRNA to target multiple mRNAs. Indeed, it has been confirmed that unlike a non-natural siRNA that targets a particular gene, each single endogenous miRNA has the potential of regulating multiple protein-coding genes, as many as 1000. On the other hand, each individual gene may be regulated by multiple miRNAs. This implies that the actions of miRNAs are

Z. Wang, *MicroRNA Interference Technologies*,
DOI: 10.1007/978-3-642-00489-6_4, © Springer-Verlag Berlin Heidelberg 2009

not gene-specific but sequence-specific as they can act on any genes carrying motifs matching their seed sites. Thus, when aiming to silence a particular gene using a naturally occurring miRNA, one may actually knockdown a group of genes. This property of miRNAs creates a hurdle for exploiting miRNA function and targets.

The RNAi pathway of siRNA-directed mRNA cleavage and the miRNA-mediated translational repression pathway are genetically and biochemically distinct. In addition to different outcomes, the two pathways have differential requirements for Paz-Piwi domain (PPD) proteins in C. elegans. Translational repression by lin-4 and let-7 depends on alg-1 and alg-2 for miRNA processing and/or stability, yet these genes are not required for RNAi (Grishok et al. 2001; Kiriakidou et al. 2007), whereas rde-1 is needed in RNAi but is not necessary for translational repression (Tabara et al. 1999). Intriguingly, miRNAs capable of translational repression can be conferred the ability to cleave targets with 3′UTRs engineered to contain completely complementary sequences (Hutvágner et al. 2002; Doench et al. 2003). Conversely, functional siRNAs can repress translation of reporter genes containing multiple imperfect binding sites in their 3′UTRs (Doench et al. 2003; Zeng et al. 2003; Martin & Caplen, 2006). The latter mimics the action of endogenous miRNAs. The findings from these elegant experiments indicate a possibility of generating non-natural artificial miRNAs simply by converting siRNAs into miRNAs.

To this end, we have developed a novel technology called microRNAs Mimics or miR-Mimics technology. The miR-Mimics generated using this approach act by miRNA mechanisms in a gene-specific manner.

4.2 Protocols

A key issue in creating functional miR-Mimics is to ensure the specificity of their interactions with their target mRNAs and to direct each interaction to discrete downstream consequences. Some key principles of this interaction have emerged based on several key studies (Doench and Sharp 2004; Lewis et al. 2003; Brennecke et al. 2005). When designing a miR-Mimic, several points should be seriously considered.

1. The key for effective miRNA action is the 5′ portion complementarity of a miRNA to its target mRNA. Complementarity of seven or more bases to residues 2–8 from the 5′ end of a miRNA, the so-called "seed site", is sufficient to confer regulation, even if the target 3′UTR contains only a single site. Sites with weaker "seed site" complementarity require compensatory pairing to the 3′ portion of the miRNA in order to confer regulation and extensive pairing to the 3′ portion of the miRNA is not sufficient to confer regulation on its own without a minimal element of 5′ complementarity;
2. The 3′ portion of the miRNA can contribute to efficiency of repression, as a modulator of suppression (Doench and Sharp 2004; Kiriakidou et al. 2004; Kloosterman et al. 2004). Thus, in addition to seed-site complementarity, appropriate 3′-end base-pairing can strengthen the effectiveness of miRNAs;

3. If efficient endonuclease cleavage is not required, then base-pairing at the site of cleavage, between bases 10 and 11, should be avoided (Elbashir et al. 2001; Haley and Zamore 2004; Martinez and Tuschl 2004);
4. To achieve efficient translational repression, multiple binding sites in the 3′UTR of a target gene may be required and for direct repression by cleavage of target mRNA, only one binding site is generally sufficient (Doench et al. 2003; Zeng et al. 2003; Doench and Sharp, 2004; Kiriakidou et al. 2004);
5. Under certain circumstances, some specific conformations, such as a bulge, in the miRNA:mRNA duplex may help in function as effective repression of the Lin-14 mRNA by the Lin-4 miRNA appears to require a bulge in the miRNA:mRNA duplex (Ha et al. 1996). The exact mechanisms are unknown, perhaps to allow the recruitment of additional RNA-binding proteins in specific contexts.

The detailed protocols are described below with the miR-Mimics for a cardiac pacemaker channel gene HCN2 and an oncogene Mdm2 as examples.

4.2.1 Designing miR-Mimics

1. Defining a unique sequence in the target gene. A fundamental requirement to be satisfied is that the 3′UTR of the target gene must contain a unique sequence distinct from other genes to elicit gene-specific action. Similar to designing a siRNA, the first step to design a miRNA mimic is to identify a stretch of sequence in the 3′UTR unique to the gene of interest (target mRNA). But unlike the full complementarity between a siRNA and its target in any region of the gene, a miRNA mimic partially base-pairs with the target sequence in the 3′UTR. The length of the sequence should be long enough for miRNA action, which is at least 8 nts and ideally >14 nts;
2. Based on the unique sequence, design a 22-nt oligonucleotides that at the 5′ portion has 8 nts (nucleotide positions 1–8 from 5′-end) fully complementary to the target mRNA;
3. To ensure the specificity of binding to the target mRNA, the oligonucleotides should have at least additional 5–6 nts complementary to the target mRNA at the 3′ portion. The base-pairing in this region may not necessarily be continuous and can be grouped into 3-nts clusters;
4. Add an "AU" to the 3′ end of the fragment. As such, a 22-nt miR-Mimic should have at least 14 complementary nucleotides to the target gene plus an "AU" overhand;
5. Design an antisense strand fully complementary to the fragment;
6. Both of the two oligonucleotide fragments need to be artificially synthesized. Many companies provide excellent services for nucleic acids research, such as Integrated DNA Technologies, Ambion, Invitrogen, etc;

7. Anneal the sense and antisense strands of the oligonucleotides to form a duplex miR-Mimic. First mix the two synthetic oligonucleotides at an equal molar concentration in an Eppendorf tube, then incubate the sample in a heat block at 95°C for 5 min and gradually cool the sample to room temperature (22–23°C);
8. Store the miR-Mimic construct at –80°C for future use;
9. Construct a negative control miR-Mimic (NC miR-Mimic) for verifying the effects and specificity of the effects of the miR-Mimic. This NC miR-Mimic should be designed based on the sequence of the miR-Mimic; simply modifying the miR-Mimic sequence to contain ~5-nt mismatches at the 5′-end "seed site". Such modified or mutated miR-Mimic is expected to lose the ability to bind to the target mRNA with sufficient affinity and is thus rendered incapable of eliciting repressive action. To form an NC miR-Mimic, the annealing procedures described above need to be repeated;
10. Synthesize an anti-miRNA inhibitor (AMO) against the miR-Mimic as an additional negative control. An AMO is designed to be an exact antisense to its target miR-Mimic. An AMO is a single-stranded oligonucleotide (ON) or oligodeoxynucleotide (ODN) fragment (see detail in Chap. 7);
11. As for SC-miRNAs, chemical modification should be done for miR-Mimic. The type of modification should be determined based on your particular need (Fig. 4.1).

4.2.2 Validating miR-Mimics

The functional activities of miR-Mimics can be verified in almost the same way as for SC-miRNAs described in Chap. 3. The detailed protocols will not be repeated in this chapter.

4.2.2.1 Luciferase Reporter Assay

Please See Chapter 3.2.2.1.

4.2.2.2 Verification of Downregulation of Target Protein

Please See Chapter 3.2.2.2.

4.2.2.3 Quantification of mRNA and miRNA

Please See Chapter 3.2.2.3.

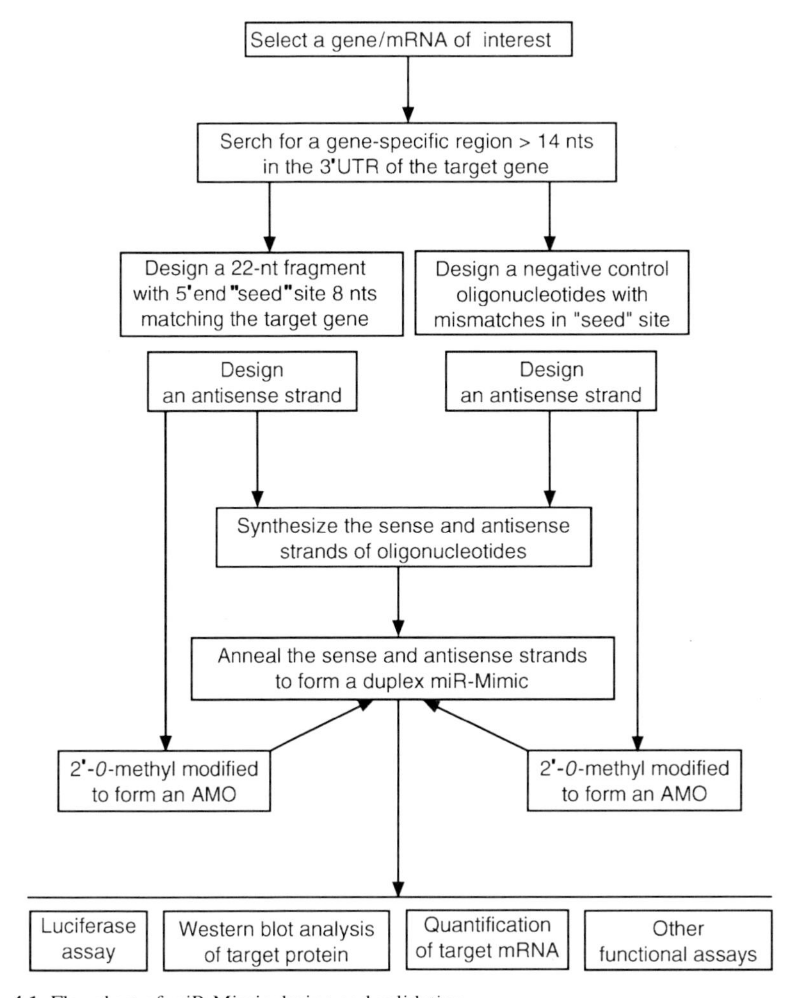

Fig. 4.1 Flowchart of miR-Mimic design and validation

4.3 Principle of Actions

The miR-Mimic technology utilizes synthetic, non-natural nucleic acids that can bind to the unique sequence of target genes (mRNAs) in a gene-specific manner and elicit post-transcriptional repressive effects in the same way as an endogenous miRNA. That is, a miR-Mimic can act only on its particular target gene but a native miRNA does can act on any gene that carries its binding sequence.

Additionally, the miR-Mimic technology produces artificial miRNAs that act by the miRNA mechanism. Thus, it will not lead to changes of expression levels of any endogenous miRNAs.

The differences among miR-Mimics and natural miRNAs and siRNAs are summarized in Table 4.1 and illustrated in Fig. 4.2.

Table 4.1 Comparisons among miR-Mimic, miRNA and siRNA

	miR-Mimic	miRNA	siRNA
Origin	Artificial, synthetic	Natural, endogenous	Artificial, synthetic
Targeting	Single mRNA	Multiple mRNAs	Single mRNA
Specificity	Gene-specific	Non-gene-specific Sequence-specific	Gene-specific
Complementarity	Partial (complementary degree upon design)	Partial (complementary degree by nature)	Full
Mechanism	Translation inhibition mRNA cleavage	Translation inhibition mRNA cleavage	mRNA cleavage

Note: miRNA and mRNA are underlined to highlight the difference

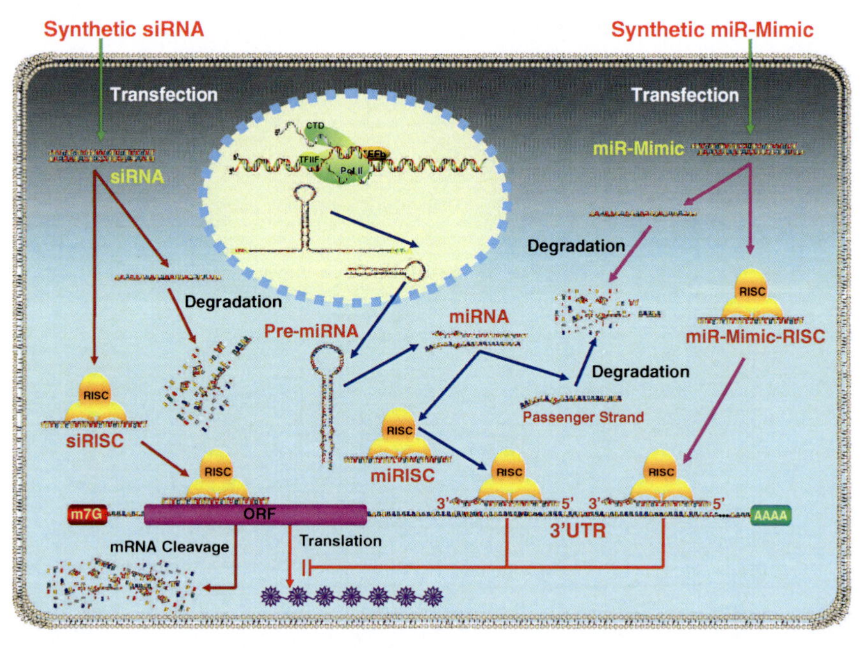

Fig. 4.2 Schematic presentation of actions of miRNA mimic (miR-Mimic) compared with the miRNA and small interference RNA (siRNA). Synthetic miR-Mimic and siRNA are introduced into the cells and endogenous miRNA is synthesized by the cell. siRNAs bind to the coding region of target miRNAs and cause mRNA cleavage; miRNAs bind to 3′UTR of multiple target mRNAs and produce non-gene-specific post-transcriptional repression to inhibit translation; miR-Mimics bind to 3′UTR of unique target mRNAs and produce gene-specific post-transcriptional repression to inhibit translation

4.4 Applications

1. To achieve gene silencing by miRNA mechanism in a gene-specific manner. As already mentioned in the previous chapter, miRNA action is not gene-specific. Thus, the gene-silencing action through SC-miRNA strategy is not gene-specific

either. The miR-Mimic technology was developed to circumvent this limitation. This property of miR-Mimics can be advantageous over SC-miRNAs when specific-genes need to be knocked down, which happens in many situations.

2. To complement the loss of endogenous miRNAs under certain conditions. In some abnormal conditions, some miRNAs are downregulated in their expression, leading to aberrantly enhanced expression of some protein-coding genes causing diseased phenotypes. Replacement of these miRNAs may reverse this process. Alternatively, application of miR-Mimics targeting the disease-causing genes to prevent their upregulation may be an efficient maneuver to tackle the problem. For example, in the development of cancer, expression of oncogenes is overexpression, partially as a result of the downregulation of their regulating miRNAs. Under such a situation, use of miR-Mimics to knock down the oncogenes may help to slow the tumorigenesis.
3. Examples of applications. We have examined the application of the technique to cardiac pacemaker genes HCN2 and HCN4 (Xiao et al. 2007). Following the protocols described above, we first identified a stretch of sequence in the 3′UTR unique to the HCN2 (or HCN4) gene that is expectedly long enough for miRNA action. Based on the unique sequence we designed a 22-nt miR-Mimic that at the 5′ end has eight nucleotides (nucleotides 2–8) and at the 3′-end has seven nucleotides, complementary to the HCN2 (or HCN4) sequence. These miR-Mimics produced substantial repression of HCN channel protein expression with concomitant depression of pacemaker activities and reduction of beating rate of cultured neonatal myocytes but with minimal effects on their mRNA levels. These cardiac automaticity-targeting miR-Mimics are expected to act like heart rate-reducing agents that have been shown to be beneficial to cardiac function during infarction and to be able to suppress ectopic beats that can be elicited by abnormal automaticity. The results demonstrated promise in utilizing the technology for gene-specific repression of expression at the protein level based on the principle of miRNA actions.

Additionally, we have also applied the miR-Mimic strategy to repress the oncoprotein Mdm2 and to suppress cancer cell growth in culture (Xiao et al. 2009).

4.5 Advantages and Problems

As an alternative to SC-miRNA technology, miR-Mimic strategy offers a couple of advantages that supplement the limitations of the former.

1. miR-Mimics can be designed to direct to solely translational process repression by sole “seed site” base-pairing or to both translational process repression and target mRNA degradation by larger degree of complementarity;
2. miR-Mimic strategy offers gene-specific targeting through miRNA mechanisms of action. This unique property makes it differ from miRNA and siRNA;

3. For a target gene, one can create many miR-Mimics at will to enhance the post-transcriptional repression, as long as the target sequence contains sufficient gene-specific stretches for designing that many miR-Mimics.

It should be noted that it is unclear yet as to what advantages the miR-Mimic technology may have over the siRNA approach. And it should also be recognized that finding gene-specific stretches of sequences restricted in the 3′UTR of protein-coding genes for designing miR-Mimics may represent a difficult task and in some cases may even be unrealistic.

References

Brennecke J, Stark A, Russell RB, Cohen SM (2005) Principles of microRNA-target recognition. PLoS Biol 3:404–418.

Doench JG, Sharp PA (2004) Specificity of microRNA target selection in translational repression. Genes Dev 18:504–511.

Doench JG, Petersen CP, Sharp PA (2003) siRNAs can function as miRNAs. Genes Dev 17:438–442.

Elbashir SM, Lendeckel W, Tuschl T (2001) RNA interference is mediated by 21- and 22-nucleotide RNAs. Genes Dev 15:188–200.

Grishok A, Pasquinelli AE, Conte D, Li N, Parrish S, Ha I, Baillie DL, Fire A, Ruvkun G, Mello CC (2001) Genes and mechanisms related to RNA interference regulate expression of the small temporal RNAs that control C. elegans developmental timing. Cell 106:23–34.

Ha I, Wightman B, Ruvkun G (1996) A bulged lin-4/lin-14 RNA duplex is sufficient for Caenorhabditis elegans lin-14 temporal gradient formation. Genes Dev 10:3041–3050.

Haley B, Zamore PD (2004) Kinetic analysis of the RNAi enzyme complex. Nat Struct Mol Biol 11:599–606.

Hutvágner G, Zamore PD (2002) A microRNA in a multiple-turnover RNAi enzyme complex. Science 297:2056–2060.

Kiriakidou M, Tan GS, Lamprinaki S, De Planell-Saguer M, Nelson PT, Mourelatos Z (2007) An mRNA m(7)G cap binding-like motif within human Ago2 represses translation. Cell 129:1141–1151.

Kloosterman WP, Wienholds E, Ketting RF, Plasterk RH (2004) Substrate requirements for let-7 function in the developing zebrafish embryo. Nucleic Acids Res 32:6284–6291.

Lewis BP, Shih IH, Jones-Rhoades MW, Bartel DP, Burge CB (2003) Prediction of mammalian microRNA targets. Cell 115:787–798.

Martin SE, Caplen NJ (2006) Mismatched siRNAs downregulate mRNAs as a function of target site location. FEBS Lett 580:3694–3698.

Martinez J, Tuschl T (2004) RISC is a 5′ phosphomonoester-producing RNA endonuclease. Genes Dev 18:975–980.

Tabara H, Sarkissian M, Kelly WG, Fleenor J, Grishok A, Timmons L, Fire A, Mello CC (1999) The rde-1 gene, RNA interference, and transposon silencing in C. elegans. Cell 99:123–132.

Xiao J, Yang B, Lin H, Lu Y, Luo X, Wang Z (2007) Novel approaches for gene-specific interference via manipulating actions of microRNAs: Examination on the pacemaker channel genes HCN2 and HCN4. J Cell Physiol 212:285–292.

Xiao J, Lin H, Luo X, Chen G, Wang H, Wang Z (2009) *miRNA-605* joins the p53 network to form a p53:*miRNA-605*:Mdm2 positive feedback loop in response to cellular stress. Nat Cell Biol (in revision).

Zeng Y, Yi R, Cullen BR (2003) MicroRNAs and small interfering RNAs can inhibit mRNA expression by similar mechanisms. Proc Natl Acad Sci USA 100:9779–9784.

Chapter 5
Multi-miRNA Hairpins and Multi-miRNA Mimics Technologies

Abstract Multi-miRNA Hairpins technology refers to a single artificial construct that can produce multiple mature miRNAs, improving gene knockdown over single miRNAs and offering expression silence of multiple genes. This technology was concurrently developed in 2006 by Zhu's laboratory (Department of Developmental and Molecular Biology, Albert Einstein College of Medicine) [Sun D, Melegari M, Sridhar S, Rogler CE, Zhu L, Biotechniques 41:59-63, 2006] and by Xu's laboratory (Department of Biochemistry and Molecular Pharmacology, University of Massachusetts Medical School) [Xia XG, Zhou H, Xu Z, Biotechniques 41:64-68, 2006a]. Similar principle was later applied by my laboratory (Department of Medicine, Montreal Heart Institute, University of Montreal) in 2008 to miR-Mimics to establish the Multi-miRNA Mimics technology that is able to silence multiple genes [Chen G, Lin H, Xiao J, Luo X, Wang Z, Biotechniques 2009]. Both Multi-miRNA Hairpins and Multi-miRNA Mimics technologies belong to the "miRNA-targeting" and "miRNA-gain-of-function" strategy. These technologies were developed based on the concept 'One-Drug, Multiple-Target' described in Sect. 2. 3.

5.1 Introduction

The SC-miRNA and miR-Mimic technologies introduced in Chaps. 3 and 4 are highly effective miRNA-gain-of-function or miRNA-targeting approaches. However, these technologies offer gain-of-function of only a singular miRNA at a time; in many circumstances, gain-of-function of multiple miRNAs is desirable (Xia et al. 2005). For instance, we may need to supply multiple miRNA tumor suppressors to suppress growth of a tumor. Use of the SC-miRNA and miR-Mimic technologies may be effective but may not be complete in terms of tumor suppression. One way to improve thorough miRNA-gain-of-function is simply to supply multiple individual SC-miRNAs or miR-Mimics. But the problems associated with this maneuver are possible unequal cellular uptake, different efficiencies of action, different

Z. Wang, *MicroRNA Interference Technologies*,

DOI: 10.1007/978-3-642-00489-6_5,

stabilities or rates of metabolism, etc., which can make control of efficacy difficult. For this reason, several new technologies have been developed in recent years to gain the power of multiple miRNA-targeting.

The method developed by Zhu's laboratory (Sun et al. 2006) takes advantage of natural miRNA hairpins existing in clusters of multiple identical or different copies (Lagos-Quintana et al. 2001; Lau et al. 2001). miRNAs in clusters suggest that polycistronic transcripts might be naturally used to enhance the efficiencies of target gene repression or to achieve linked multigene repression. The idea is that polycistronic transcripts could be generated artificially to achieve better knockdown and linked multi-gene knockdown by modified miRNAs.

A recent advance in gene silencing methodology is to modify the stem sequence of pre-miR-30 hairpin structure to achieve knockdown of artificially targeted genes (Zeng et al. 2002; Zhou et al. 2005; McManus et al. 2002; Boden et al. 2004; Stegmeier et al. 2005; Silva et al. 2005; Dickins et al. 2005).

5.2 Protocols

5.2.1 Construction of Multi-miRNA Hairpins

5.2.1.1 Lentivirus-CMV-GFP Method (Zhu's Protocols) (Sun et al. 2006)

1. Select a miRNA for your study and select the hairpin precursor sequence of that miRNA from miRBase (Griffiths-Jones, 2006; http://microrna.sanger.ac.uk/registry/; the Wellcome Trust Sanger Institute);
2. Designing hairpin units. To ensure efficient processing of the hairpin, add flanking sequences to both the 5′-end and 3′-end of the precursor miRNA. The length of the flanking sequence can vary from 10 to 150 nts; It was reported that the pre-miR-30 hairpin can be generated correctly and efficiently in vitro when it is flanked by a minimum of 22 nts of its natural flanking sequences at its 5′ side and 15 nts at its 3′ side (Zeng and Cullen 2005; Zeng et al. 2005; Lee et al. 2003). In the study reported by Sun et al. (2006), a 118-nt extension was used for pre-miR-30.

Then add a short artificial sequence containing restriction enzyme sites to both 5′- and 3′-ends of the flanking-fragment-carrying precursor. As many appropriate and needed precursor units can be linked together but each unit should carry a unique set of restriction enzyme sites. For example, three hairpin units with four different restriction enzyme sites (*Xba*I, *Spe*I, *Bam*HI, and *Kpn*I for a lentivirus vector) were used in the study reported by Sun et al. (2006). The first unit carries 5′-*Xba*I and 3′-*Kpn*I, the second unit has 5′-*Kpn*I and 3′-*Spe*I and the third unit contains 5′-*Spe*I and 3′-*Bam*HI.

Thus, each of the hairpin units carries a pre-miRNA, the extension portions and the artificial linkers with restriction enzyme sites.

3. Synthesis of hairpin units. Synthesize the hairpin units using the service provided by companies or by PCR method in your own laboratory;
4. Construction of Multi-miRNA Hairpins. Link the precursor units to form a Multi-miRNA Hairpins structure by mixing all the hairpin units (Note that because of the restriction enzyme sites carried in the units, the units are able to connect with each other according to their enzyme sites);
5. Construction of lentivirus vector. Insert the multi-hairpin constructs into *Xba*I and *Bam*HI sites of lenti-CMV-GFP about 70 nt downstream from a cytomegalovirus (CMV; i.e., promoter), followed by a GFP open reading frame (ORF) to mark hairpin-expressing cells (Follenzi et al. 2002; Sun et al. 2006). Lentiviruses can achieve high-efficiency gene transduction in both dividing and quiescent cells;
6. Functional tests. Deliver the effect of the lentiviruses into cells by infection and test the target-gene knockdown effects of the multi-miRNA hairpins using luciferase activity assay and Western blot analysis. Alternatively, the multi-hairpin constructs can be delivered into cells by transfection using lipid carriers without being cloned into a lentivirus vector.

Note: (1) The total length of each hairpin unit should not exceed 100 nts. (2) There should be an inverse relationship between the intensity of GFP fluorescence and hairpin copy numbers. Since processing of hairpins leads to destruction of the primary transcript from which GFP is translated, this inverse relationship may suggest that the multi-hairpin transcripts are processed more efficiently than the single-hairpin transcript. Sun et al. (2006) confirmed that the amounts of one-, two-, and three-hairpin transcripts are all significantly reduced compared with the amount of the hairpin-less GFP transcript. (3) Most strikingly, knockdown efficiency of a single hairpin can be significantly improved by the presence of a second hairpin even when it does not target the same gene. Indeed, the degree of mature small RNA production and gene knockdown efficiency achieved by the two-hairpin construct over the single-hairpin construct is clearly disproportionably larger than one fold increase in hairpin copy number, suggesting that the addition of a second hairpin provides robust stimulation to the miRNA processing process. (4) The multiple hairpins in the construct are all functionally active and the relative position of a hairpin in the multi-hairpin cluster is flexible, not restricted by a particular order (Fig. 5.1).

5.2.1.2 Tetracycline-Regulated Pol II Promoter Method (Xu's Protocols) (Xia et al. 2006a)

1. Construction of tetracycline-regulatable human ubiquitin C (UbC) promoters. Insert the tetracycline-responsive element (TRE) into two sites: one within the promoter region and the other in the first noncoding exon of the human UbC gene. For this, introduce a restriction site for XhoI at the position 5 nucleotides (nts) downstream of the TATA box by changing the sequence 5′-…TATATAAGGACGCGCCGG…-3′ to 5′…TATATAAGGACGCtCgaG… -3′. Then, to insert the TRE in the first exon, create an XhoI site at the position 26 nts downstream from the transcription start site by changing the sequence

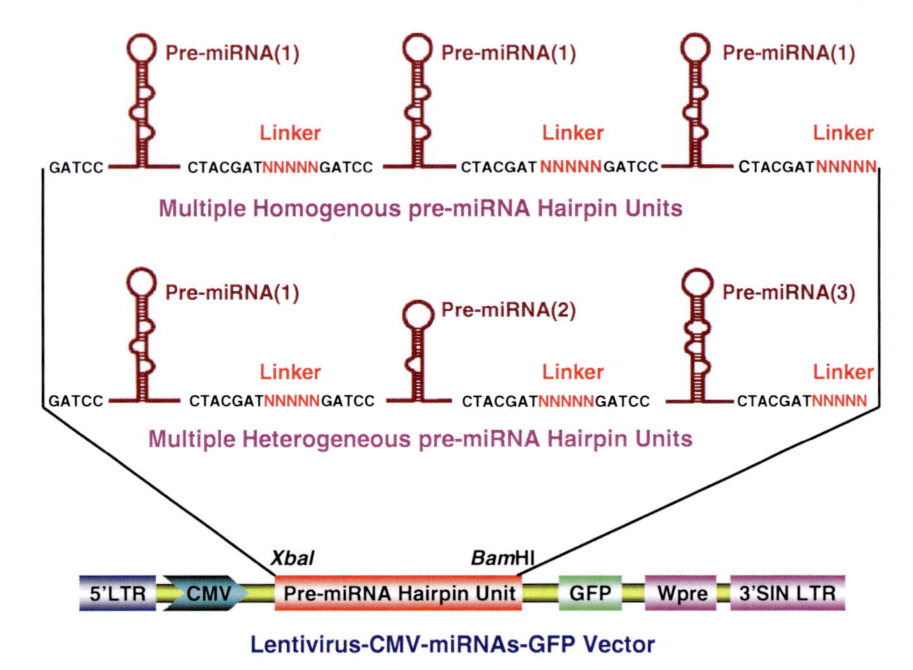

Fig. 5.1 Schematic illustration of design of multi-miRNA hairpin vector based primarily on the study reported by Sun et al. (2006), showing the arrangement of a multiple hairpin structure containing three identical (upper) or three different (lower) pre-miRNAs. "NNNNNN" accommodates various restriction enzyme sites as needed. The hairping unit structure is cloned into lentivirus vector to form a lentivirus-CMV-miRNAs-GFP vector

5′-…GGGTCGCGGTTC TT…-3′ to 5′-…GGGTCtCGagTCTT…-3′. One or two TRE sites (5′-TCCCTATCAGTGATAGAGA-3′) are then inserted at the XhoI site;

2. Construction of a self-regulatable UbC vector (Shi 2003). PCR clone a coding sequence for tetracycline-controlled transcriptional silencer (tTS) from pTeTtTS vector (Clontech Laboratories, Mountain View, CA, USA). Then replace tTS for the green fluorescent protein (EGFP)-coding sequence of the UbC-EGFP vector. Digest the newly constructed UbC-tTS vector sequentially with restriction enzymes *Xho*I and *Not*I. Digest the pIRES2-EGFP vector (Clontech Laboratories) with *Xho*I and *Not*I to release the internal ribosome entry site (IRES) of the encephalomyocarditis virus and the EGFP coding region. Finally, ligate the IRES-EGFP into the UbC-tTS vector to generate a UbC-tTS-IRES-EGFP vector;
3. Chemically synthesize the fifth intron of the human actin gene with a 5-nt flanking sequence at each end. Then, insert the fragment at the *Eco*RI site, which is 10 nts from the 3′-end of the first intron of the human ubiquitin C gene in the UbC-TRE4 vector. The *Eco*RV site will be destroyed by this insertion;

4. Chemically synthesize the fourth intron of human actin gene with a 5-nt flanking sequence. Then, insert the fragment at the *Xho*I site (inactivated by insertion) downstream of the EGFP coding region;
5. Select target genes and define gene-specific sequence for study. Chemically synthesize the individual hairpins targeting the selected genes. In a study reported by Xia et al., (2006a), the hairpins were designed to mimic human miR-30 that targets the human superoxide dismutase-1 (sod1);
6. Then, insert these hairpins sequentially into the *Eco*RI site within the first intron of the human UbC gene, the *Pst*I site (introduced during the chemical synthesis) in the fifth intron of the human actin gene and the *Xho*I site (also introduced during the chemical synthesis) inside the fourth intron of the human actin gene;
7. Finally, for functional test, construct a target vector based on the pGL2-luc vector (Promega, Madison, WI, USA), which synthesizes firefly luciferase. Insert synthetic oligonucleotides containing the target sequences 100 bp downstream from the firefly luciferase opening reading frame (ORF) in the 3′UTR. All constructs should be sequence verified (Fig. 5.2).

Note: (1) Human UbC promoter is a ubiquitous promoter that works well in transgenic animals (Li et al. 2003; Lois et al. 2002; Xia et al. 2006b). By incorporating TRE into the promoter, it becomes regulatable by doxycycline. Because UbC drives the expression of tTS, one expects that the expression level of EGFP should be low in the absence of doxycycline for the construct UbC-TRE-tTS-IRES-EGFP, which has a TRE in the first exon. By addition of doxycycline, the expression will be induced. (2) Xia et al. (2006a) compared the construct that carries hairpins against all three different protein-coding genes (human *sod1*, mouse *Sod2*, or mouse *Dj-1*) with that carry only one hairpin for one specific target. They found that the degree of silencing is comparable between the single and multiple hairpin constructs. (3) Xia et al. (2006a) confirmed that their multi-hairpin construct can silence multiple targets simultaneously in the same cells. (4) According to the inventors of the method (Xia et al. 2006a), a tetracycline-regulated pol II promoter construct allows the expression of up to only three miRNA hairpins.

5.2.2 Construction of Multi-miRNA Mimics (Chen et al. 2009)

While the above approaches utilized hairpins that can generate a particular canonical miRNA (miR-30), these methods can also be applied to engineer multiple hairpins that can produce artificial miR-Mimics (Xiao et al. 2007) to give rise to gene-specific miRNA actions. Taking this advantage, my laboratory developed Multi-miR-Mimic technology for this purpose (Chen et al. 2009).

1. Defining unique motifs in the target gene. Follow the same protocols for designing miR-Mimics, the first step to design a multi-miRNA mimic is to identify a few stretches of sequence in the 3′UTR of a gene that are unique/specific to

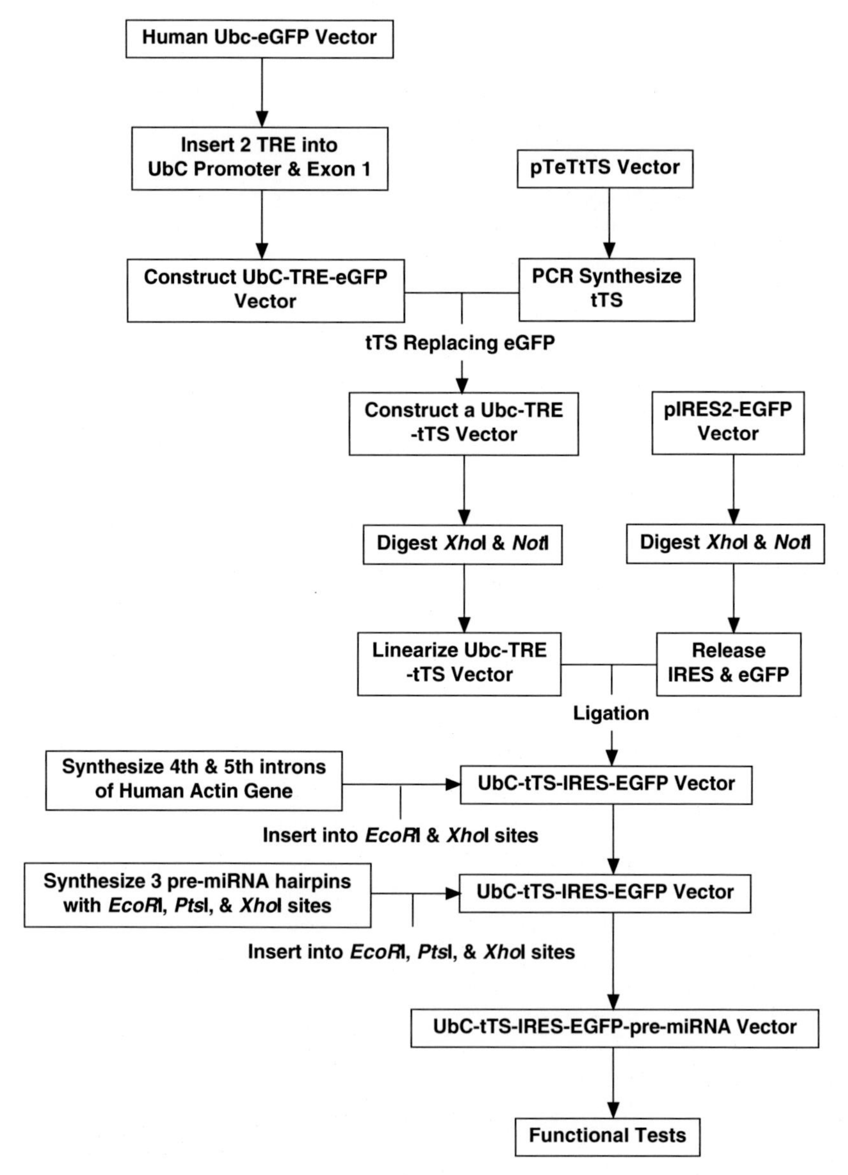

Fig. 5.2 Flowchart for designing a tetracycline-regulated Pol II promoter based on the study reported by Xiao et al. (2006)

the gene of interest (target mRNA) for gene-specific targeting. Alternatively, if multiple genes are needed to be interfered with simultaneously, define the gene-specific sequences in the 3′UTRs of the selected target genes and design multi-miRNA mimics for these genes;

2. Based on the unique motifs, design 22-nt oligonucleotide fragments that at the 5′-portion has 8 nts (nucleotide positions 1 to 8 from 5′-end) fully complementary to the target mRNA;

3. To ensure the specificity of binding to the target mRNA, the oligonucleotides should have at least additional 5–6 nts complementary to the target mRNA at the 3′-portion. The base-pairing in this region may not necessarily be continuous and can be grouped into 3-nts clusters;
4. Then insert each of these fragments (miR-Mimics) into the pre-miR-30 by replacing the mature miR-30 sequence. You then have chimeric fragments each carrying a miR-Mimic and a pre-miR-30 sequence minus mature miR-30;
5. Follow the procedures (4)–(6) in Sect. 5.1.1 to construct multi-miR-Mimic hairpins in a lentivirus-CMV-GFP vector;
6. Test the efficacy of the construct in using a luciferase activity reporter vector.

A simpler way to construct a Multi-miR-Mimic is to link multiple individual miR-Mimics together by a linker sequence (TGUUUTUGUTUGUTUUUTUGGC), which does not contain any miRNA binding sites. But the disadvantage of this simple approach is a less intracellular stability of constructs.

Note: (1) The beauty of our Multi-miR-Mimics approach is to use miR-30-less pre-miR-30 hairpin structure to cheat the cells to process it to generate "mature" miR-Mimics. (2) The gene-specific action, which may be highly desirable in many cases, is another beauty of our Multi-miR-Mimics approach. (3) Like the Lentivirus-CMV-GFP method and the Tetracycline-regulated Pol II promoter method, the Multi-miR-Mimics method can host either multiple homogeneous miR-Mimics to generate identical miR-Mimics for amplifying actions on a particular gene or multiple heterogeneous miR-Mimics to generate distinct miR-Mimics for diverse actions on multiple target genes.

5.3 Principle of Actions

The ideas behind the Multi-miRNA Hairpins and Multi-miRNA Mimics technologies are quite straightforward. They all take advantage of expression vehicles that are able to give high efficiency and high fidelity of producing multiple mature miRNAs or artificial miR-Mimics for any particular purpose, providing both an enhanced single-gene knockdown and linked multi-gene knockdowns (Table 5.1).

5.4 Applications

1. In theory, these multi-silencing constructs can be used in investigations where knockdown of multiple genes is required both for investigating whether genes are in sequential or parallel pathways and whether genes are performing redundant roles and acting synergistically or antagonistically and for targeting any pathological processes involving multiple factors.
2. These multiple knockdown strategies can also be used in transgenic mice. These technologies should improve the use of gene knockdown in laboratory research

Table 5.1 Comparisons among SC-miRNA, multi-miR-Hairpins and multi-miR-Mimic

	SC-miRNA	Multi-miR-Hairpins	Multi-miR-Mimics
Origin	Natural, endogenous	Natural, endogenous	Artificial, synthetic
miRNA	Single miRNA	Multiple miRNAs	Multiple miR-Mimics
Specificity	Non-gene-specific Sequence-specific	Non-gene-specific Sequence-specific	Gene-specific
Complementarity	Partial (complementary degrees by nature)	Partial (complementary degrees by nature)	Partial (complementary degrees as designed)
Mechanism	Translation inhibition or mRNA cleavage, depending on complementarity	Translation inhibition or mRNA cleavage, depending on complementarity	Translation inhibition or mRNA cleavage, as designed

Note: miRNA and mRNA are underlined to highlight the difference

and facilitate the development of more efficacious gene silencing-based therapeutics.

3. Examples of applications. The Lentivirus-CMV-GFP method has been applied to knockdown several genes, S-phase kinase protein 2 (SKP2), cyclin-dependent kinase 2 (Cdk2) and androgen receptor (AR) (Sun et al. 2006). SKP2 is an F-box protein targeting cell-cycle regulators including cycle-dependent kinase inhibitor p27KiP1 via ubiquitin-mediated degradation and is frequently overexpressed in a variety of cancer cells and has been implicated in oncogenesis Cdk2 in maintaining and regulating cell cycle kinetics.

The Tetracycline-regulated Pol II promoter method has been used to target Dj-1, human sod1 (soperoxide dismutase 1) and mouse sod2 genes (Xia et al. 2006a). Dj-1 is a ubiquitous protein that was first described as an oncogene and later found to be associated with monogenic Parkinson's disease. Sod1 and sod2 encode the endogenous antioxidant enzyme.

The Multi-miRNA Mimics method has been used to target antiapoptotic oncogenic transcription factors E2F, NF-κB and Mdm2 (Chen et al. 2009).

5.5 Advantages and Limitations

1. The systems described in this chapter can achieve more effective miRNA-targeting or knockdown of protein-coding genes than the conventional gene-silencing methods, such as siRNA or antisense.
2. Knockdown efficiency of a single hairpin can be significantly improved by the presence of a second hairpin even when it does not target the same gene or even when it does not target any genes.

3. Compared with the gene knockout approach in animals, the multiple miRNA hairpin approaches can avoid having to carry out multiple crosses of individual knockout lines, which is difficult if not impossible and, therefore, can accelerate investigations of gene-gene interactions in vivo.

References

Boden D, Pusch O, Silbermann R, Lee F, Tucker L, Ramratnam B (2004) Enhanced gene silencing of HIV-1 specific siRNA using microRNA designed hairpins. Nucleic Acids Res 32:1154–1158.

Dickins RA, Hemann MT, Zilfou JT, Simpson DR, Ibarra I, Hannon GJ, Lowe SW (2005) Probing tumor phenotypes using stable and regulated synthetic microRNAs precursors. Nat Genet 37:1289–1295.

Follenzi A, Sabatino G, Lombardo A, Boccaccio C, Naldini L (2002) Efficient gene delivery and targeted expression to hepatocytes in vivo by improved lentiviral vectors. Hum Gene Ther 13:243–260.

Griffiths-Jones S (2006) miRBase: The microRNA sequence database. Methods Mol Biol 342:29–138.

Lagos-Quintana M, Rauhut R, Lendeckel W, Tuschl T (2001) Identification of novel genes coding for small expressed RNAs. Science 294:853–858.

Lau NC, Lim LP, Weinstein EG, Bartel DP (2001) An abundant class of tiny RNAs with probable regulatory roles in *Caenorhabditis elegans*. Science 294:858–862.

Lee Y, Ahn C, Han J, Choi H, Kim J, Yim J, Lee J, Provost P, Rådmark O, Kim S, Kim VN (2003) The nuclear RNase III Drosha initiates microRNA processing. Nature 425:415–419.

Li X, Makela S, Streng T, Santti R, Poutanen M (2003) Phenotype characteristics of transgenic male mice expressing human aromatase under ubiquitin C promoter. J. Steroid Biochem Mol Biol 86:469–476.

Chen G, Lin H, Xiao J, Luo X, Wang Z (2009) Multi-miRNA-Mimics strategy offers a powerful and diverse gain-of-function of miRNAs for gene silencing. Biotechniques.

Lois C, Hong EJ, Pease S, Brown EJ, Baltimore D (2002) Germline transmission and tissue-specific expression of transgenes delivered by lentiviral vectors. Science 295:868–872.

McManus MT, Petersen CP, Haines BB, Chen J, Sharp PA (2002) Gene silencing using microRNA designed hairpins. RNA 8:842–850.

Shi Y (2003) Mammalian RNAi for the masses. Trends Genet 19:9–12.

Silva JM, Li MZ, Chang K, Ge W, Golding MC, Rickles RJ, Siolas D, Hu G, Xu Z (2005) Second-generation shRNA libraries covering the mouse and human genomes. Nat Genet 37:1281–1288.

Stegmeier F, Hu G, Rickles RJ, Hannon GJ, Elledge SJ (2005) A lentiviral microRNA-based system for single-copy polymerase II-regulated RNA interference in mammalian cells. Proc Natl Acad Sci USA 102:13212–13217.

Sun D, Melegari M, Sridhar S, Rogler CE, Zhu L (2006) Multi-miRNA hairpin method that improves gene knockdown efficiency and provides linked multi-gene knockdown. Biotechniques 41:59–63.

Xia XG, Zhou H, Xu ZS (2005) Promises and challenges in developing RNAi as a research tool and therapy for neurodegenerative diseases. Neurodegenerative Diseases 2:220–231.

Xia XG, Zhou H, Xu Z (2006a) Multiple shRNAs expressed by an inducible pol II promoter can knock down the expression of multiple target genes. Biotechniques 41:64–68.

Xia XG, Zhou H, Samper E, Melov S, Xu Z (2006b) Pol II-expressed shRNA knocks down Sod2 gene expression and causes phenotypes of the gene knockout in mice. PLoS Genetics 2:e10.

Xiao J, Yang B, Lin H, Lu Y, Luo X, Wang Z (2007) Novel approaches for gene-specific interference via manipulating actions of microRNAs: Examination on the pacemaker channel genes HCN2 and HCN4. J Cell Physiol 212:285–292.
Zeng Y, Cullen BR (2005a) Efficient processing of primary microRNA hairpins by Drosha requires flanking nonstructured RNA sequences. J Biol Chem 280:27595–27603.
Zeng Y, Cai X, Cullen BR (2005b) Use of RNA polymerase II to transcribe artificial microRNAs. Methods Enzymol 392:371–380.
Zeng Y, Wagner EJ, Cullen BR (2002) Both natural and designed micro RNAs can inhibit the expression of cognate mRNAs when expressed in human cells. Mol Cell 9:1327–1333.
Zhou H, Xia XG, Xu Z (2005) An RNA polymerase II construct synthesizes short-hairpin RNA with a quantitative indicator and mediates highly efficient RNAi. Nucleic Acids Res 33:e62.

Chapter 6
miRNA Transgene Technology

Abstract The miRNA-targeting or miRNA-gain-of-function technologies introduced in Chaps. 3–5 can only create transient forced expression of miRNAs in in vitro conditions, which may be inadequate when the research or therapy requires long-lasting, stable overexpression of miRNAs for in vivo evaluation of function consequent to miRNA-targeting. Transgenic mice models have been used to tackle the problem. There are in general two different approaches for establishing miRNA transgene animals. One is the conventional transgene approach that has been widely used for protein-coding genes. This approach was first applied to miRNA research in 2006 by Costinean et al. [Proc Natl Acad Sci USA 103:7024–7029, 2006] and Peng et al. [Proc Natl Acad Sci USA 103:2252–2256, 2006], then in 2007 by Lu et al. [Dev Biol 310:442–453, 2007], and recently by us as well Zhang et al. [Science, 2009]. The second is a creative approach utilizing artificial intronic miRNAs generated by inserting a transposon into the intron of a protein-coding gene, which was originally developed by Ying's laboratory in 2003 [Lin SL, Chang D, Wu DY, Ying SY, Biochem Biophys Res Commun 310:754–760, 2003; Lin SL, Chang SJ, Ying SY, Methods Mol Biol 342:321–334, 2006]. The miRNA-transgene technology belongs to the "miRNA-targeting" and "miRNA-gain-of-function" strategy and is primarily used for studying miRNA target genes, cellular function and pathological role in animal models. This strategy is largely based on the concept of 'miRNA as a Regulator of a Cellular Function' introduced in Sect. 2.1.2.

6.1 Introduction

The miRNA-targeting or miRNA-gain-of-function technologies introduced in the preceding chapters primarily create transient overexpression of miRNAs in an in vitro environment and may be inadequate to address the aspects where the research or therapy requires long-lasting overexpression of miRNAs and/or in vivo evaluation of function consequent to miRNA-targeting. To circumvent the obstacle, in vivo animal models with stable expression of miRNAs of interest are needed. Transgenic

Z. Wang, *MicroRNA Interference Technologies*,
DOI: 10.1007/978-3-642-00489-6_6,

mice meet this need and have been in use for the expression of "foreign" genes for decades. The technology to generate such mice is well established and the use of genetically engineered mice has allowed researchers to explore fundamental functions of genes in a mammal that shares substantial similarities with human physiology and pathology. Genetically engineered mice are often used as animal models of human diseases and are vital tools in investigating disease development and in developing and testing novel therapies. Gene targeting in embryonic stem cells allows endogenous genes to be specifically altered.

As knowledge regarding precise genetic abnormalities underlying a variety of dermatological conditions continues to emerge, the ability to introduce corresponding alterations in endogenous gene loci in mice, often at a single base pair level, has become essential for detailed studies of these genetic diseases. In theory, miRNAs can also be expressed in the same way as protein-coding genes to achieve gene-silencing effects.

Indeed, recent studies have shown that short RNAs could be engineered into introns (Ying and Lin 2006). This could easily translate into an effective laboratory tool for elucidation of miRNAs and repression of particular sets of genes. The approach may also find its use as a regulator in engineered pathways in synthetic biology and gene therapy. Ying's group was the first to introduce the transgene approach into miRNA research. In their elegant study published in 2003 (Lin et al. 2003), they reported the development of a state-of-the-art transgenic strategy for silencing specific genes in zebrafish, chicken and mice, using intronic miRNAs. By insertion of a hairpin pre-miRNA structure into the intron region of a gene, they made successful transcription of mature miRNAs by RNA polymerase II, co-expressed with a protein-coding gene transcript and efficient excision out of the encoding gene transcript by natural RNA splicing and processing mechanisms. In conjunction with retroviral transfection systems, the hairpin pre-miRNA construct was further inserted into the intron of a cellular gene for tissue-specific expression regulated by the gene promoter.

In addition to the above artificial intronic miRNA approach, the conventional transgene approach has also been used as a miRNA-targeting strategy in miRNA research (Costinean et al. 2006; Peng et al. 2006; Lu et al. 2007; Zhang et al.2008).

A detailed description on the procedures for generating miRNA transgene mouse models is provided below.

6.2 Protocols

6.2.1 Conventional Transgene Methods

To generate miRNA transgene mice, the general procedures for protein-coding genes apply. These involve the following sequential steps: (1) construction of a vector carrying the stem-loop pre-miRNA, (2) introduction of the vector into fertilized mouse eggs, (3) egg transplantation into female mice and (4) verification of transgene in the offspring. Successful and efficient delivery of the constructed vector into eggs is a key step.

There are a number of methods for vector delivery: (1) micro-injecting the vector into pronuclei of eggs; (2) introducing the construct into one-cell embryos through lentiviral infection or direct injection of the viruses (Rubinson et al. 2003). However, this method requires a high-titer viral stock, which is not easy to obtain. Retroviral vectors will not work as transgenes since they will be silenced in mouse embryos; and (3) transfecting mouse embryonic stem cells with a miRNA expression cassette along with a drug selection marker via electroporation or lentiviral infection (Everett et al. 1996; Peli et al. 1996; Kunath et al. 2003; Chen et al. 2006). Transfected clones are selected through the drug resistance. The expression of the mature miRNA is then analyzed to identify clones with the desired construct. miRNA-transgene mice can then be generated via blastocyst injection or tetraploid aggregation.

6.2.1.1 Vector construction

1. Select a miRNA for your study. For example, we studied mmu-miR-328 transgene (Zhang et al. 2009), in a study reported by (Costinean et al. 2006), mmu-miR-155 transgene was created and in another study, the investigators established mmu-miR-17–92 cluster Transgene (Lu et al. 2007);
2. PCR synthesize a fragment of around 300 bp long containing the precursor miRNA (e.g., pre-miR-328) sequence using the mouse genomic DNA as a template;
3. Subclone the fragment into the *Sal*I and *Hind*III sites of Bluescript vector (Promega) carrying the cardiac-specific α-myosin heavy chain (αMHC) promoter and human growth hormone poly(A)$^{+}$signal (Fig. 6.1).

(Alternatively, other plasmids can be used, such as the pBSVE6BK (pEμ) plasmid containing the Eμ nhancer V_H promoter for Ig heavy chains, alongside the 3′UTR and the poly(A) of the human β-globin gene (Costinean et al. 2006) and pKC4 plasmid containing the mouse 6.5 kb Sftpc promoter (Lu et al. 2007). Peng et al. 2006) used a transgenic shRNA vector (pTshRNA) based on the original pSUPER (Brummelkamp et al. 2002), in which the shRNA expression is driven by H1 promoter. In pTshRNA vector, the 3′-end (~1 kb) of the *H1* gene was included to help the expression of shRNA. In addition, an EGFP cassette was added for easy genotyping of the transgenic animals);

4. Digest the plasmid at the *Spe*I site to release the pre-miR-328 sequence flanked by 5′-end αMHC promoter and 3′-end poly(A);
5. Separate the fragment on an agarose gel and purify it by phenol/chloroform extraction;
6. Dialyze the fragment for 36 h against 150-ml injection buffer (0.1 mM EDTA, 5 mM Tris-HCl). Then filter the fragment through a 0.22-µm filter. Prepare the DNA sample at a concentration of 3 ng/µl ready for injection;

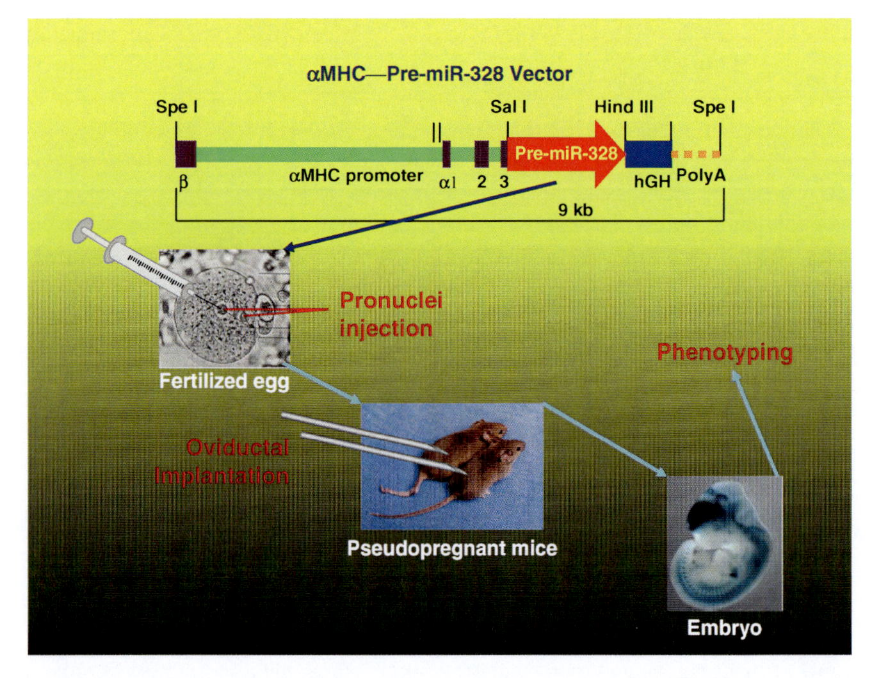

Fig. 6.1 Schematic illustration of engineering the vector carrying pre-miR-328 for generating transgenic mice by pronuclear injection based on the study by Zhang et al. (2009). A fragment (350 bp) containing pre-miR-328 sequence was PCR amplified from the mouse genomic DNA. The fragment was subcloned into the Sal I and Hind III sites of Bluescript vector (Promega) carrying the cardiac-specific α myosin heavy chain (αMHC) promoter and human growth hormone poly(A)$^+$ signal. The plasmid was digested at the Spe I site to release the pre-miR-328 sequence flanked by 5′-end αMHC promoter and 3′-end poly(A). The DNA fragment was individually micro-injected into the pronuclei of mouse one-cell embryos to generate heterozygous miR-328 transgene mice

6.2.1.2 Microinjection

7. Egg production for injections. Superovulate sexually immature female mice (4–5 weeks of age) by consecutive premenstrual syndrome (PMS) hormone and human chorionic gonadotropin (HCG) hormone injections to obtain sufficient quantity (>250) of eggs for injection. Let these female mice be mated with stud males immediately following the HCG injection;
8. Harvesting eggs. Harvest the eggs the next day from the ampulla of the oviduct of the mated females and treat them with hyaluronidase to remove nurse cells. Store the fertilized eggs in M16 media (37°C, 5% CO_2) until injection;
9. Injecting eggs. Remove 30–50 eggs at a time from the incubator for injection. Micro-inject each of the eggs individually with the prepared DNA fragment under a high magnification microscope. After microinjection, return the eggs to the incubator. Remove the eggs that do not survive injection;

6.2.1.3 Egg Implantation

10. Prepare pseudo-pregnant female mice by mating with the vasectomized males;
11. On the day of micro-injection, anesthetize the pseudo-pregnant females with 1% pentobarbital (16 mg/kg) and open the abdomen. Implant the injected eggs in a group of 10–15 bilaterally into the oviduct of these mice. Allow the animals to recover from anesthesia on a warming plate and then place them back to the animal room. The animals must be kept under sterile conditions throughout their pregnancy;

6.2.1.4 Verification of Transgene

12. Extract the genomic DNA from the tail of the transgenic mice;
13. Screen pups for the presence of the transgene by Southern blot on tail-extracted genomic DNA;
14. Study transgenic hemizygous mice and compare with their wild-type counterparts. Genotype mice by PCR amplification on tail-extracted DNA. For mmu-miR-328 in Bluescript vector, one can use the forward primer recognizing αMHC (5250–5268): 5′-CCTTACCCCACATAGACCT-3′ and the reverse primer for miR-328 (58–39): 5′-CTGTAGATACTTTCTCCCT-3′;
15. Northern blot analysis of transgenic offspring on total RNA extracted from the tissues under test, to establish transgenic lines. For example, cardiac tissue was used in our study, spleen was used in the study reported by (Costinean et al. 2006) and lung was used by (Lu et al. 2007).

The whole procedure for generating miRNA transgene mouse models are illustrated in Fig. 6.2.

6.2.2 Artificial Intronic miRNA Methods

The key to successful generation of miRNA transgene lines using the Artificial Intronic miRNA methods is to construct an artificial splicing-competent intron (SpRNAi) mimicking the natural structure of a pre-mRNA intron (Kramer 1996; Mougin et al. 2002; Fairbrother et al. 2002). The SpRNAi consists of consensus nucleotide elements representing: splice donor and acceptor sites, a branch-point domain, a poly-pyrimidine tract and linkers for insertion into gene constructs. Additionally, an insert sequence that is either homologous or complementary to a targeted exon is located within the artificial intron between the splice donor site and the branch-point domain. This portion of the intron would normally represent a region unrecognized by small nuclear ribonucleoproteins (snRNPs) during RNA splicing and processing. To facilitate the accuracy of RNA splicing, the SpRNAi should also contain a translation stop codon in its 3′-proximal region, which if present

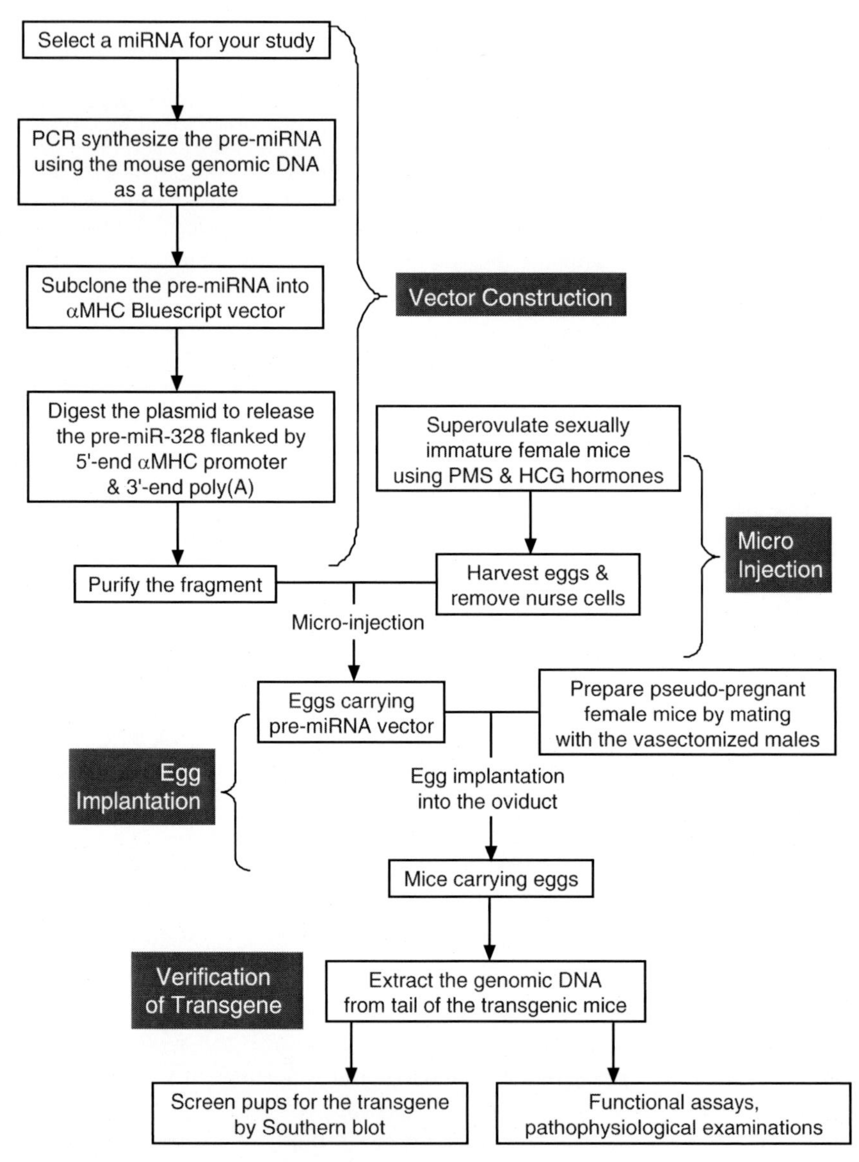

Fig. 6.2 Flowchart for generating miRNA transgene mouse model based on the study reported by Zhang et al. (2008)

in a cytoplasmic mRNA would signal the diversion of the defective pre-mRNA to the nonsense-mediated mRNA decay (NMD) pathway. For intracellular expression of the SpRNAi, we further inserted the above construct into a red fluorescent protein (rGFP) gene. Because the intronic insertion disrupts the expression of functional rGFP, it becomes possible to determine the occurrence of intron splicing and

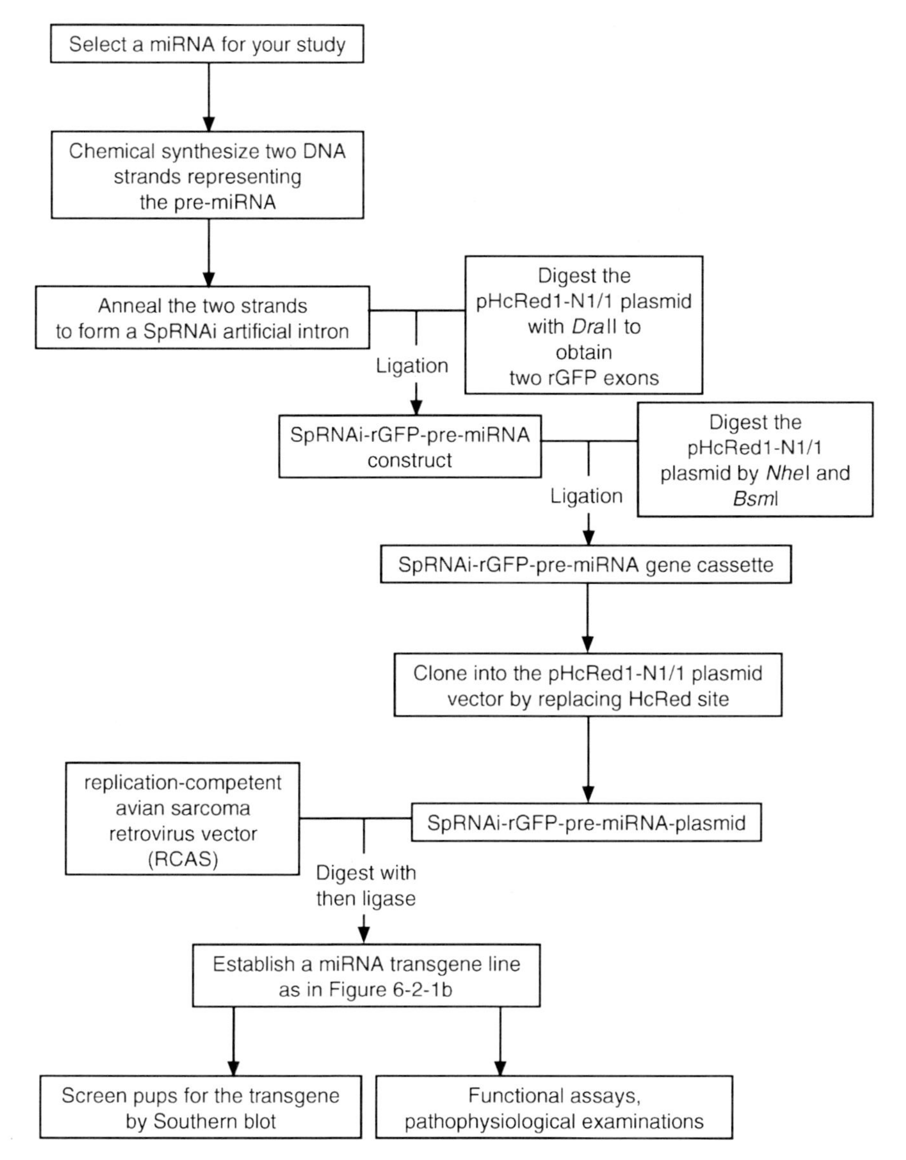

Fig. 6.3 Flowchart for generating miRNA transgene mouse model based on the study reported in the journal articles (Lin et al. (2003, 2006; Lin and Ying 2006a) and a book chapter (Lin and Ying 2006b). SpRNAi: an artificial splicing-competent intron that can mimic the natural structure of a pre-mRNA intron

rGFP-mRNA maturation through the reappearance of red fluorescent emission in transfected cells (Figs. 6.3 and 6.4).

Readers are referred to the detailed step-by-step protocols for generating miRNA transgene lines using the Artificial Intronic miRNA methods published by the

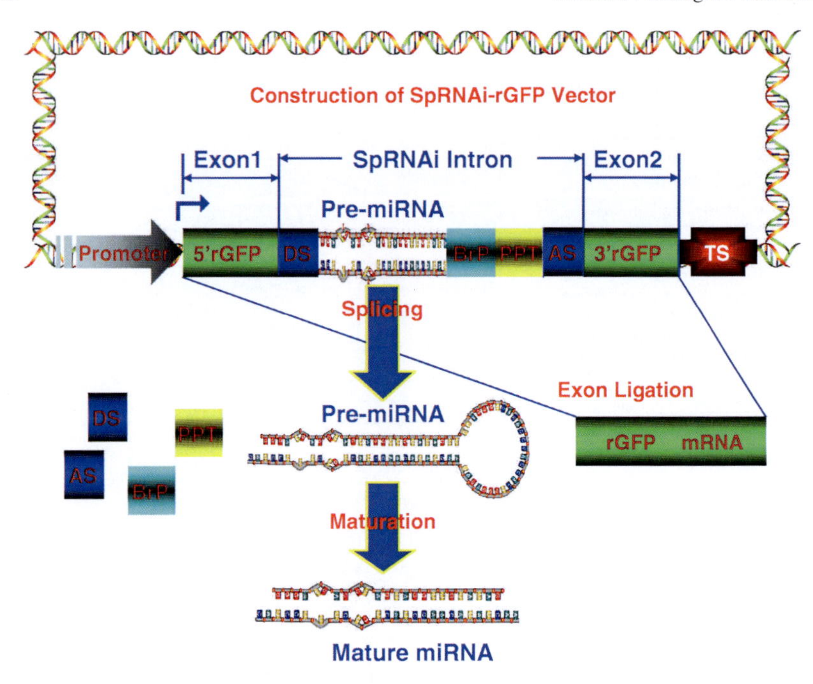

Fig. 6.4 Strategy for constructing intronic miRNA transgene using artificial SpRNAi-rGFP-gene vector, based on the study reported by Lin and Ying (2006b). The SpRNAi-rGFP vector is composed of a promoter, an artificial SpRNAi intron flanked by red fluorescent protein exon1 (5′rGFP) and exon2 (3′rGFP) and a translation termination codon (TS). The SpRNAi intron contains a 5′-splice donor site (DS), a 3′-splice acceptor site (AS), a poly-pyrimidine tract (PPT), a branch-point domain (BrP) and an oligomer insert representing a pre-miRNA. During mRNA maturation, the pre-miRNA spliced out of the SpRNAi intron and further processed into mature miRNA by Dicer

inventors of the technology as journal articles (Lin et al. 2003, 2006; Lin and Ying 2006a) and a book chapter (Lin and Ying 2006b).

6.2.2.1 Construction of SpRNAi-rGFP Plasmid Vector

1. A SpRNAi-rGFP artificial intron is formed by hybridization of a sense SpRNAi sequence and its exact antisense. Sense SpRNAi sequence: 5′-dephosphorylated GTAAGTGGTC CGATCGTCGC GACGCGTCAT TACTAACACTAT CAT-ACTTATC CTGTCCCTTT TTTTTCCACA GCTAGGACCT TCGTGCA-3′; Antisense SpRNAi sequence: 5′-dephosphorylated TGCACGAAGG TCCTAG-CTGT GGAAAAAAAA GGGACAAGGAT AAGTATGATA GTTAGTAATG ACGCGTCGCG ACGATCGGAC CACTTAC-3′.
2. Synthesize the two fragments using the commercial services. Then anneal the two strands in a hybridization buffer (in mM: 100 KOAc, HEPEAS-KOH,

MgOAC, pH7.4), by first heating to 94°C for 3 min followed by cooling down to 65°C for 10 min to form **a SpRNAi artificial intron.**

3. Digest the pHcRed1-N1/1 plasmid with *Dra*II to obtain two rGFP exons containing an AG-GN nucleotide break with 5′-G(T/A)C protruding nucleotides at the cleaved ends.
4. Blunt one end of the rGFP exons to remove the 5′-GTC protruding nucleotides using T4 DNA polymerase. Keep the 5′-GAC protruding end intact in the other rGFP exon.
5. Ligate the SpRNAi intron with the 5′-GAC-rGFP exon at the 3′-*Dra*II-restricted ends.
6. Purify the ligation product in 1% agarose gel to obtain the SpRNAi-rGFP construct.
7. Digest the pHcRed1-N1/1 plasmid by *Nhe*I and *Bsm*I.
8. Purify the digested sequences in 1% agarose gel.
9. Ligate the SpRNAi-rGFP and the pHcRed1-N1/1 digested by NheI/BsmI.
10. Purify the ligation product in 1% agarose gel to obtain the final SpRNAi-rGFP gene cassette.
11. Clone the SpRNAi-rGFP gene cassette into the pHcRed1-N1/1 plasmid vector by replacing the original HcRed site, in order for the SpRNAi-rGFP gene to be able to express in transfected cells. The cloning is set at the XhoI and XbaI restriction sites of both the pRNAi-rGFP gene cassette and the pHcRed1-N1/1 plasmid vector.
12. After purification of the ligation product to acquire the SpRNAi-rGFP plasmid vector, amplify the plasmid using DH5α transformation-competent *E. coli* cells.
13. Recover the SpRNAi-rGFP **plasmid** in autoclaved ddH$_2$O.

6.2.2.2 Insertion of Pre-miRNA Hairpin into the SpRNAi-rGFP Plasmid

14. Select a miRNA of your interest for your study and pull out the sequence of the pre-miRNA. Add the restriction sites of PuvI (GTCCG) and Mlu1 (ATGAC) to the 5′- and 3′-ends of the sense and its antisense strands.
15. Synthesize the sense and the antisense strands of the selected pre-miRNA in the form of oligodeoxyribonucleotides or DNA using commercial services. Anneal the two strands in the hybridization buffer described above.
16. Digest the SpRNAi-rGFP plasmid and the pre-miRNA construct, respectively. Then purify the digested sequences in 1% agarose gel.
17. Ligate the digested SpRNAi-rGFP plasmid and the pre-miRNA construct to acquire the SpRNAi-rGFP-pre-miRNA plasmid.
18. Amplify the ligation construct in DH5α transformation-competent *E. coli* cells.
19. Recover the plasmid in autoclaved ddH$_2$O.
20. For in vitro studies in cell lines, the SpRNAi-rGFP-pre-miRNA plasmid can be transfected into cells in culture by liposome reagents.

6.2.2.3 Construction of Retroviral Vector Carrying the SpRNAi-rGFP-pre-miRNA Plasmid

21. Purchase the replication-competent avian sarcoma virus (RCAS) vector.
22. Digest the SpRNAi-rGFP-pre-miRNA plasmid and the RCAS retroviral vector, separately. Purify the digested sequences in 1% agarose gel.
23. Ligate the two digested constructs and purify them.
24. Follow steps 7–15 in section 12.2.1 (Conventional Transgene Methods) to establish a miRNA transgene line.

6.2.3 Cre-loxP Knock-in Methods

To better analyze miRNA and target gene function, it is desirable to establish miRNA transgene in a temporally and spatially controllable manner. Cre-loxP system offers a handy means to conditionally activate or inactivate miRNAi in an irreversible manner (Coumoul et al. 2004; Kasim et al., 2003, 2004; Tiscornia et al. 2004; Ventura et al. 2004).

In the conditional activation system (Cre-on), a loxP-flanked random stuffer sequence is inserted between the Pol III promoter and the miRNA-encoding sequence. On the introduction of Cre recombinase, the recombination of the two loxP sites excises the stuffer sequence, allowing miRNA expression.

In the conditional inactivation system (Cre-off), a Pol III-miRNA expression cassette along with an EGFP marker sequence is flanked by loxP sites. In the absence of Cre, the miRNA and the adjacent EGFP marker can be expressed constitutively. Upon the introduction of Cre, the miRNA cassette and the EGFP marker are removed by Cre excision, leading to the termination of miRNA expression.

One can take advantage of the availability of various transgenic mouse lines that express Cre conditionally to generate spatiotemporally controlled miRNA transgene models. This approach has been used to conditionally express miR-155 and an enhanced green fluorescent protein (EGFP) reporter in mature B cells, in a Cre-dependent manner (Thai et al. 2007) and to generate mice carrying the miR-155 knock-in and the CD21-Cre alleles (Fig. 6.5).

6.3 Principle of Actions

Two different approaches of miRNA-transgene models are based upon different mechanisms for overexpression of mature miRNAs as desired but give rise to the same effects (gene silencing) by the same principle of actions (partial base-pairing through miRISC). Thus, I will focus on the different mechanisms for overexpression between the conventional transgene methods and the artificial intronic miRNA methods.

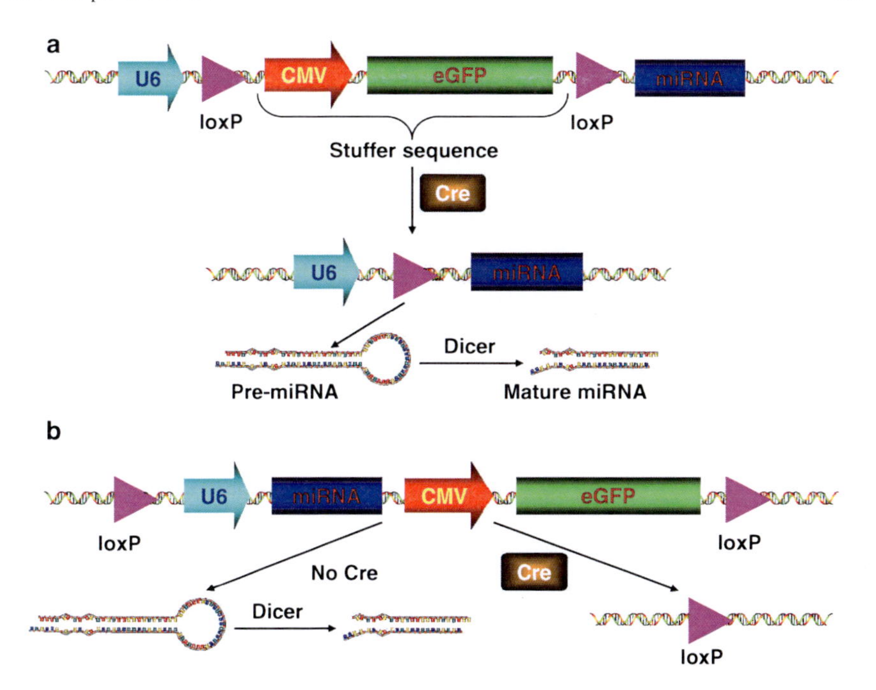

Fig. 6.5 Conditional RNAi strategies using the Cre-loxP system modified from Gao and Zhang (2007). (**a**) In the conditional activation (Cre-on) approach, the pre-miRNA-encoding sequence and its promoter are separated by a buffer sequence flanked with two loxP sites. On the introduction of Cre, the buffer sequence is excised and pre-RNA is expressed. (**b**) In the conditional inactivation system (Cre-off), the pre-miRNA cassette along with an eGFP marker sequence is flanked by two loxP sites. In the absence of Cre, the pre-miRNA as well as the adjacent eGFP marker is expressed constitutively. Upon the introduction of Cre, both pre-miRNA cassette and the eGFP marker are removed and the miRNAi effect is terminated

6.3.1 Conventional Transgene Methods

Expression of mature miRNAs in the transgene mice is driven by the promoter incorporated in the vector construct. This provides an opportunity to make tissue-specific expression of the miRNA; for example, αMHC confers cardiac-specific expression.

6.3.2 Artificial Intronic miRNA Methods

In humans, nearly 97% of the genome is non-coding DNA, which varies from one individual to another and changes in these sequences are frequently noted to manifest clinical and circumstantial malfunction. For instance, type 2 myotonic dystrophy and fragile X syndrome were found to be associated with miRNAs derived from introns. Intronic miRNA is a new class of miRNAs derived from the processing of non-protein-coding regions of gene transcripts (Lin et al. 2006a, b). The intronic miRNAs differ from previously described intergenic miRNAs in the requirement of RNA polymerase (Pol)-II and spliceosomal components for its biogenesis. We performed an analysis of the miRNAs registered in miRBase and identified 45% of them as intronic miRNAs, indicating that nature favors expression of intronic miRNAs. This provides an advantage for creating the artificial intronic miRNAs for stable, endogenous expression and the associated gene silencing.

When a transposon is inserted in the intron, it becomes an intronic miRNA, taking advantage of the protein synthesis machinery, i.e., mRNA transcription and splicing, as a means for processing and maturation. Indeed, (Lin et al., 2003, 2006b) confirmed that by packaging up human spliceosome-recognition sites along with an exonic insert into an artificial intron, the splicing and processing of such an exon-containing intron in either sense or antisense conformation produced equivalent gene silencing effects, while a palindromic hairpin insert containing both sense and antisense strands resulted in synergistic effects. Compared to the Conventional Transgene Methods, the Artificial Intronic miRNA Methods make use of the endogenous host gene promoters.

Moreover, the rGFP protein encoded by the SpRNAi-rGFP-pre-miRNA vector can serve as a quantitative marker for measuring the titer levels of the pre-miRNA-expressing RCAS virus using either flow cytometry or the lentivirus quantification kit.

6.4 Applications

1. To achieve long-lasting gain-of-function of a target miRNA for studying developmental issues and disease progression. The transgene effect is literally permanent to an animal and can usually be transmitted down through generations. This offers a platform for long-term investigation of development-related issues and pathogenesis involving remodeling processes.
2. To allow for controllable or conditional overexpression of the target miRNA in relation to particular pathophysiological states.
3. To study miRNA-gain-of-function in in vivo whole-animal context. Compared to the miRNA-gain-of-function approaches introduced in Chaps. 3–5, one obvious application of the transgenic model is to study miRNA functions in whole-animal contexts.
4. Examples of applications. The miRNA transgene animal models have been used for miRNA research in a number of occasions. And the outcomes from these studies support a wide range of applications of the approach to understanding miRNA function, target genes and pathological roles.

Lu et al. (2007) generated miR-17–92 cluster transgenic embryos that maintain a high expression level in the epithelium throughout development. Using the model, they were able to discover that high levels of miR-17–92 cluster help to maintain lung progenitor cells in a proliferative and undifferentiated state. They also indentified Rbl2 as a novel target gene for miR-17–5p, which provides a mechanistic link between the miRNAs and the hyperproliferative growth in lung progenitor cells.

Peng et al. (2006) designed a transgenic shRNA vector (pTshRNA) based on the original pSUPER and created a transgene mouse line using this vector. In a small-scale screening for developmental defects in the kidney using the transgene model, they uncovered a potential role of *Id4* in the formation of the renal medulla.

Transgenic mice with overexpression of miR-155 develop a lymphoproliferative disease resembling the human diseases, thus strongly suggesting that miR-155 is directly implicated in the initiation and/or progression of these diseases (Costinean et al. 2006). The disease is, for the most part, polyclonal, suggesting that only the overexpression of miR-155 or additional few genetic changes is sufficient to induce malignancy. Because malignancies are for the most part monoclonal, this finding suggests that miR-155 could be the downstream target of signal transduction pathways activated in cancer. This is direct evidence that overexpression of a miRNA results in the development of a neoplastic disease, highlighting their potential role in human malignancies. Interestingly, we observed overexpression of miR-155 in solid tumors such as breast, lung and colon cancer (in lung cancers, overexpression of miR-155 was an indicator of bad prognosis) (Volinia et al. 2006). The Eμ-mmu-miR-155 transgenic mouse will also be a useful tool to devise new therapeutic approaches to treat different forms of acute lymphoblastic leukemia or high-grade lymphomas in humans.

We recently used the conventional transgene model to elucidate the key role of miR-328 in controlling experimental atrial fibrillation (Zhang et al. 2009). In this study, we were also able to identify an array of gene encoding proteins critical to cardiac intracellular Ca^{2+} handling, including L-type Ca^{2+} channel α1c subunit (CACNA1C), L-type Ca^{2+} channel β1 subunit (CACNB1), Na^{+}/Ca^{2+} exchanger (NCX1), Ca^{2+}-ATPase (SERCA2) and Ca^{2+}/calmodulin-depedent protein kinase (CAMK2D) and ppp3CA/B (calcineurin).

The Artificial Intronic miRNA technology has been developed for in vitro evaluation of gene function, in vivo gene therapy and generation of transgenic animal models (Lin and Ying 2006a). Intron-derived miRNA is not only able to induce RNAi in mammalian cells but also in fish, chicken embryos and adult mice, demonstrating the evolutionary preservation of this gene regulation system in vivo. These miRNA-mediated animal models provide useful means to reproduce the mechanisms of miRNA-induced disease in vivo and will shed further light on miRNA-related therapies.

6.5 Advantages and Limitations

The discovery that miRNAs can silence gene expression in mammalian cells has revolutionized biomedical research. The most successful application of miRNAi technologies has been to study gene function in cultured human or mouse cells.

However, the knockdown effect of miRNA is only transient in these conditions. To achieve a more sustained gene-silencing effect, hairpin RNA expressed from a vector is preferred. An additional benefit is that miRNAi can now be applied in vivo through delivering miRNA-expressing vectors by transgenic technology. Transgenic miRNAi not only allows the study of biological processes not present in cultured cells but also offers chronic therapeutic potentials. The transgene models offer enormous advantages in miRNA research and miRNAi applications, as outlined below.

1. The development of transgenic technologies in mice has allowed us to manipulate the mouse genome and to study the consequences of specific genetic alterations on a global perspective and in in vivo system. Generated transgenic embryos that maintain a high expression level of a miRNA can be transmitted for many generations in these transgenic animals for studying developmental and remodeling processes.
2. Gao and Zhang (2007) stated "The beauty of transgenic miRNAi is that, unlike gene knockouts, one does not have to breed the transgene into homozygosity to see the effect." In fact, one can even observe phenotypic consequences of the miRNAi effect on transgenic embryos or mice at the F_0 generation. Thus miRNA transgenes behave like dominant-negative alleles of the genes of interest. This feature, combined with the relatively inexpensive pronuclear injection technique, makes it possible to perform genetic screenings in mice. For developmental phenotypes, the screening can be done with little mouse cage costs, since there is no breeding needed when the injected one-cell embryos can be scored directly after they have developed to certain stages. Obviously, the screening cannot have as high throughput as in lower organisms, such as *C. elegans* (Fraser et al. 2000; Sonnichsen et al. 2005), since the constructs have to be injected one by one. However, it is high throughput comparing with other mutagenesis strategies in mice.
3. The transgenic models can be made inducible and ultimately allow both temporal and spatial gene inactivation and are highly informative for studying a variety of miRNA-related biological and pathological processes.
4. The transgenic animal models of human diseases serve as a front line for drug evaluation. Transgenic animal models are valuable tools for testing gene functions and drug mechanisms in vivo, providing reliable models for the pre-clinical approval of therapy for human disease to develop more efficient compounds and functional antibodies. They are also the best similitude of a human body for etiological and pathological research of diseases. All pharmaceutically developed drugs must be proven safe and effective in animals before approval by the Food and Drug Administration to be used in clinical trials.

References

Brummelkamp TR, Bernards R, Agami R (2002) A system for stable expression of short interfering RNAs in mammalian cells. Science 296:550–553.

Chen Y, Chen H, Hoffmann A, Cool DR, Diz DI, Chappell MC, Chen A, Morris M (2006) Adenovirus-mediated small-interference RNA for in vivo silencing of angiotensin AT1a receptors in mouse brain. Hypertension 47:230–237.

Choi WY, Giraldez AJ, Schier AF (2007) Target protectors reveal dampening and balancing of Nodal agonist and antagonist by miR-430. Science 318:271–274.

Coumoul X, Li W, Wang RH, Deng C (2004) Inducible suppression of Fgfr2 and Survivin in ES cells using a combination of the RNA interference (RNAi) and the Cre-LoxP system. Nucleic Acids Res 32:e85.

Costinean S, Zanesi N, Pekarsky Y, Tili E, Volinia S, Heerema N, Croce CM (2006) Pre-B cell proliferation and lymphoblastic leukemia/high-grade lymphoma in E(mu)-miR155 transgenic mice. Proc Natl Acad Sci USA 103:7024–7029.

Everett CA, West JD (1996) The influence of ploidy on the distribution of cells in chimaeric mouse blastocysts. Zygote 4:59–66.

Fairbrother WG, Yeh RF, Sharp PA, Burge CB (2002) Predictive identification of exonic splicing enhancers in human genes. Science 297:1007–1013.

Fraser AG, Kamath RS, Zipperlen P, Martinez-Campos M, Sohrmann M, Ahringer J (2000) Functional genomic analysis of C. elegans chromosome I by systematic RNA interference. Nature 408:325–330.

Gao X, Zhang P (2007) Transgenic RNA Interference in Mice. Physiology 22:161–166.

Kasim V, Miyagishi M, Taira K (2004) Control of siRNA expression using the Cre-loxP recombination system. Nucleic Acids Res 32:e66.

Kasim V, Miyagishi M, Taira K (2003) Control of siRNA expression utilizing Cre-loxP recombination system. Nucleic Acids Res Suppl 3:255–256.

Kramer A (1996) The structure and function of proteins involved in mammalian pre-mRNA splicing. Annu Rev Biochem 65:367–409.

Kunath T, Gish G, Lickert H, Jones N, Pawson T, Rossant J (2003) Transgenic RNA interference in ES cell-derived embryos recapitulates a genetic null phenotype. Nat Biotechnol 21:559–561.

Lin SL, Chang D, Wu DY, Ying SY (2003) A novel RNA splicing-mediated gene silencing mechanism potential for genome evolution. Biochem Biophys Res Commun 310:754–760.

Lin SL, Chang SJ, Ying SY (2006) Transgene-like animal models using intronic microRNAs. Methods Mol Biol 342:321–334.

Lin SL, Kim H, Ying SY (2008) Intron-mediated RNA interference and microRNA (miRNA). Front Biosci 13:2216–2230.

Lin SL, Ying SY (2006a) Gene silencing in vitro and in vivo using intronic microRNAs. Methods Mol Biol 342:295–312.

Lin SL, Ying SY (2006b) Gene slicing using intronic microRNAs. In *MicroRNA Protocols*, ed by Ying SY, Humana Press. pp 305–320.

Lu Y, Thomson JM, Wong HY, Hammond SM, Hogan BL (2007) Transgenic over-expression of the microRNA miR-17–92 cluster promotes proliferation and inhibits differentiation of lung epithelial progenitor cells. Dev Biol 310:442–453.

Mougin A, Gottschalk A, Fabrizio P, Luhrmann R, Branlant C (2002) Direct probing of RNA structure and RNA–protein interactions in purified HeLa cell's and yeast spliceosomal U4/U6.U5 Tri-snRNP particles. J Mol Biol 317:631–649.

Peli J, Schmoll F, Laurincik J, Brem G, Schellander K (1996) Comparison of aggregation and injection techniques in producing chimeras with embryonic stem cells in mice. Theriogenology 45:833–842.

Peng S, York JP, Zhang P (2006) A transgenic approach for RNA interference-based genetic screening in mice. Proc Natl Acad Sci USA 103:2252–2256.

Rubinson DA, Dillon CP, Kwiatkowski AV, Sievers C, Yang L, Kopinja J, Rooney DL, Ihrig MM, McManus MT, Gertler FB, Scott ML, Van Parijs L (2003) A lentivirus-based system to functionally silence genes in primary mammalian cells, stem cells and transgenic mice by RNA interference. Nat Genet 33:401–406.

Sonnichsen B, Koski LB, Walsh A, Marschall P, Neumann B, Brehm M, Alleaume AM, Artelt J, Bettencourt P, Cassin E, Hewitson M, Holz C, Khan M, Lazik S, Martin C, Nitzsche B, Ruer M, Stamford J, Winzi M, Heinkel R, Roder M, Finell J, Hantsch H, Jones SJ, Jones M, Piano F, Gunsalus KC, Oegema K, Gonczy P, Coulson A, Hyman AA, Echeverri CJ (2005) Full-genome RNAi profiling of early embryogenesis in Caenorhabditis elegans. Nature 434:462–469.

Thai TH, Calado DP, Casola S, Ansel KM, Xiao C, Xue Y, Murphy A, Frendewey D, Valenzuela D, Kutok JL, Schmidt-Supprian M, Rajewsky N, Yancopoulos G, Rao A, Rajewsky K (2007) Regulation of the germinal center response by microRNA-155. Science 316:604–608.

Tiscornia G, Tergaonkar V, Galimi F, Verma IM (2004) CRE recombinase-inducible RNA interference mediated by lentiviral vectors. Proc Natl Acad Sci USA 101:7347–7351.

Ventura A, Meissner A, Dillon CP, McManus M, Sharp PA, Van Parijs L, Jaenisch R, Jacks T (2004) Cre-lox-regulated conditional RNA interference from transgenes. Proc Natl Acad Sci USA 101:10380–10385.

Volinia S, Calin GA, Liu CG, Ambs S, Cimmino A, Petrocca F, Visone R, Iorio M, Roldo C, Ferracin M, Prueitt RL, Yanaihara N, Lanza G, Scarpa A, Vecchione A, Negrini M, Harris CC, Croce CM (2006) A microRNA expression signature of human solid tumors defines cancer gene targets. Proc Natl Acad Sci USA 103:2257–2261.

Xiao J, Yang B, Lin H, Lu Y, Luo X, Wang Z (2007) Novel approaches for gene-specific interference via manipulating actions of microRNAs: Examination on the pacemaker channel genes HCN2 and HCN4. J Cell Physiol 212:285–292.

Ying SY, Chang DC, Lin SL (2008) The microRNA (miRNA): Overview of the RNA genes that modulate gene function. Mol Biotechnol 38:257–268.

Ying SY, Lin SL (2004) Intron-derived microRNAs–fine tuning of gene functions. Gene 342:25–28.

Ying SY, Lin SL (2006) Current perspectives in intronic micro RNAs (miRNAs). J Biomed Sci 13:5–15.

Zhang Y, Lu Y, Wang N, Lin H, Pan Z, Gao X, Zhang F, Zhang Y, Xiao J, Shan H, Luo X, Chen G, Qiao G, Wang Z, Yang B (2009) Control of experimental atrial fibrillation by microRNA-328. Science (in revision).

Chapter 7
Anti-miRNA Antisense Oligonucleotides Technology

Abstract One of the indispensable approaches in miRNA research as well as in miRNA therapy is inhibition or loss-of-function of miRNAs. Multiple steps in pathway for miRNA biogenesis could be targeted for inhibition of miRNA production and maturation. Thus far, nonetheless, the anti-miRNA antisense inhibitor oligoribonucleotides (AMO) technology used to target mature miRNAs has found most value for these applications. Standard AMO is a single-stranded 2′-O-methyl (2′-OMe)-modified oligoribonucleotide fragment exactly antisense to its target miRNA. The AMO technology was initially established in 2004 by Tuschl's laboratory (Laboratory of RNA Molecular Biology, The Rockefeller University) [Meister G, Landthaler M, Dorsett Y, Tuschl T, RNA 10:544–550, 2004] and by Zamore's laboratory (Department of Biochemistry and Molecular Pharmacology, University of Massachusetts Medical School) [Hutvágner G, Simard MJ, Mello CC, Zamore PD, PLoS Biol 2:465–475, 2004]. The idea was however originated from Boutla et al. (Nucleic Acids Res 31:4973–4980, 2003) who used antisense 2′-deoxyoligonucleotides to sequence-specifically inactivate miRNAs in microinjected *D. melanogaster* embryos. Since then, AMO technology has undergone many important modifications to enhance the efficiency and specificity of miRNA interference. These include cholesterol moiety-conjugated 2′-OMe modified AMOs called antagomiR [Krutzfeldt J, Rajewsky N, Braich R, Rajeev KG, Tuschl T, Manoharan M, Stoffel M, Nature 438:685–689, 2005], locked nucleic acid (LNA)-modified AMOs [Ørom UA, Kauppinen S, Lund AH, Gene 372:137–141, 2006; Davis S, Lollo B, Freier S, Esau C, Nucleic Acids Res 34:2294–2304, 2006], 2′-O-methoxyethyl (2′-MOE) [Esau C, Davis S, Murray SF, Yu XX, Pandey SK, Pear M, Watts L, Booten SL, Graham M, McKay R, Subramaniam A, Propp S, Lollo BA, Freier S, Bennett CF, Bhanot S, Monia BP, Cell Metab 3:87–98, 2006; Davis S, Lollo B, Freier S, Esau C, Nucleic Acids Res 34:2294–2304, 2006], 2′-flouro (2′-F) (Davis et al. 2006), phosphorothioate backbone modification and peptide nucleic acid (PNA)-modified AMOs [Fabani MM, Gait MJ, RNA 14:336–346, 2008]. The AMO technology belongs to the "targeting-miRNA" and "miRNA-loss-of-function" strategy. For the sake of clarity and convenience, I here designate all different types

Z. Wang, *MicroRNA Interference Technologies*,
DOI: 10.1007/978-3-642-00489-6_7, © Springer-Verlag Berlin Heidelberg 2009

of anti-miRNA antisense AMO and the conventional anti-mRNA antisense ASO, though ASO has also been used to refer to anti-miRNA antisense by some authors.

7.1 Introduction

Multiple steps along the path of biogenesis of miRNAs can be interfered to achieve miRNA knockdown. Targeted degradation of pri-miRNA transcripts in the nucleus with RNaseH-based antisense oligodeoxynucleotide molecules offers an opportunity of inhibiting production of multiple miRNAs from a polycistronic pri-miRNA transcript. RNaseH recognizes RNA–DNA duplexes, cleaving the RNA strand (Wu et al. 2004).

Inhibition of Drosha processing of the pri-miRNA with antisense oligodeoxynucleotides that do not support RNaseH activity has been reported in zebrafish (Kloosterman et al. 2007).

Antagonizing the endoribonuclease Dicer to block miRNA maturation using antisense ODNs may be another feasible approach. Indeed, the efficacy of knocking down miRNAs by destructing Dicer has been documented (Murchison et al. 2005; Chen et al. 2008).

Alternative to the antisense strategy, RNAi technology may be applied for the same purpose. However, one consideration for these strategies is that the timing of inhibition of miRNA function may be delayed, as existing mature miRNA levels must decay before inhibition will be observed. Several studies have suggested that mature miRNA turnover is slow (Lee et al. 2003, 2006; Schaefer et al. 2007). Therefore, these may not be effective strategies for short-term in vitro studies.

The base-pair interaction between miRNAs and mRNAs is essential for the function of miRNAs; therefore, a logical approach of silencing miRNAs is to use a nucleic acid that is antisense to the miRNA. These anti-miRNA oligonucleotides (AMOs) specifically and stoichiometrically bind and efficiently and irreversibly silence their target miRNAs. This AMO approach has been used in numerous studies to identify many cellular functions of miRNAs (Boutla et al. 2003; Meister et al. 2004; Hutvágner et al. 2004; Davis et al. 2006; Ørom et al. 2006; Hammond 2006; Weiler et al. 2006; Esau 2008) and has been considered a plausible strategy for miRNA therapy of human disease (Hammond 2006; Weiler et al. 2006; Wang et al. 2008). It is the most straightforward and apparently most effective strategy tested so far to block the function of miRNAs in RISC. It mediates potent and miRNA-specific inhibition of miRNA function, providing a powerful "loss-of-function" strategy for interfering miRNA expression.

Among the various forms of AMOs, antagomiR that has its end conjugated with a cholesterol moiety has demonstrated most impressive effectiveness against target miRNAs, intracellular stability, particularly for in vivo applications (Krutzfeldt et al. 2005). This particular form of AMO is discussed in more detail in a later section of this chapter.

7.2 Protocols

7.2.1 Designing AMOs

Designing AMOs is in general straightforward, involving only a few simple steps. Below is a description of designing a standard AMO: 2′-OMe-modified anti-miRNA antisense.

1. Select a miRNA of interest for your study. Obtain the mature sequence of the miRNA from miRBase (http://microrna.sanger.ac.uk/registry/; the Wellcome Trust Sanger Institute). Important: obtain the guide strand and discard the passenger strand;
2. Design an oligonucleotide fragment (RNA) exactly complementary (base-pairing or matching) to the guide strand of the mature miRNA (Fig. 7.1). Note that it should be an RNA fragment. At micromolar concentration, antisense oligodeoxynucleotides (ODNs) may block miRNA function in *Drosophila* embryos (Boutla et al. 2003) but the poor stability of DNA oligonucleotides in vivo may limit their utility;
3. Chemically synthesize the designed oligoribonucleotide fragment with 2′-O-methyl modification; this is your AMO. You may attach a C7 amino linker to the 3′-end of AMO if you plan to introduce post-synthetic conjugation of non-nucleotidic residues such as biotin succinimidyl esters for monitoring transfection. In case you synthesize AMOs using your own equipment, the methods described by Meister et al. (2004) can be followed. 2′-O-Methyl oligoribonucleotides can be synthesized using 5′-DMT, 2′-O-methyl phosphoramidites (Proligo) on 1-μM synthesis columns loaded with 3′-aminomodifier (TFA) C7 ICAA control pore glass support (Chemgenes). The aminolinker is added to also use the oligonucleotides for conjugation to amino group reactive reagents. The synthesis products are deprotected for 16 h at 55°C in 30% aqueous ammonia and then precipitated by the addition of 12 ml absolute 1-butanol. The full-length product is then gel-purified using a denaturing 20% polyacrylamide gel;
4. For negative control, synthesize an oligonucleotide fragment complementary to the passenger strand of the miRNA;
5. Transfection of AMOs (10–50 nM) in cultured cells or in tissues by in vivo gene transfer techniques. If using cationic lipid, one generally cannot transfect more than 100 nM AMO of any chemistry without non-specific toxicity.

AMOs designed in this way have been shown to be highly effective in silencing miRNA function; it has been consistently documented in numerous studies that AMOs are able to wipe out the level of the targeted miRNAs to virtually zero (less than 0.1% of control levels).

However, a study reported by Vermeulen et al. (2007) claimed that extending the AMO sequences attaching flanking elements dramatically increases overall potency. This study demonstrated that the length and composition of sequences surrounding an AMO have considerable effects on overall inhibitor potency. Increasing AMO

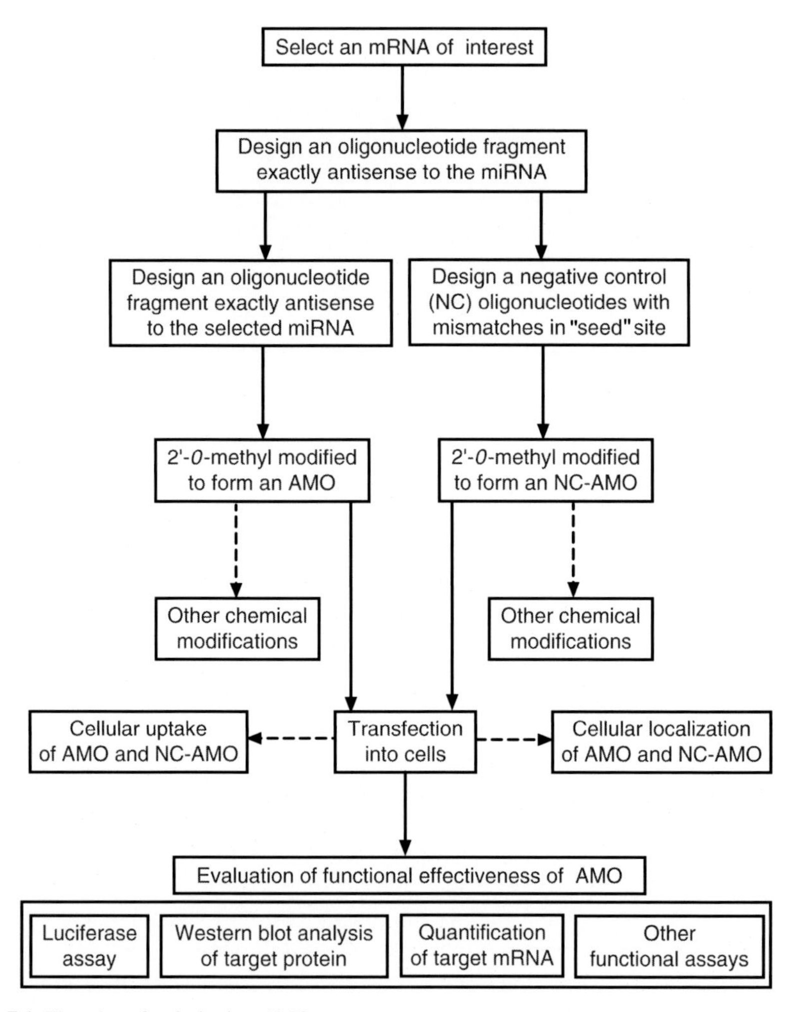

Fig. 7.1 Flowchart for designing AMO

length by adding as many as 16 nt to each side of the miRNA-complementary core increased AMO potency as much as tenfold. Incorporation of highly structured, double-stranded flanking regions around an AMO significantly increases inhibitor function and allows for miRNA inhibition at subnanomolar concentrations. AMO molecules having symmetrical flanking regions of >10-nt can inhibit target miRNAs to a much greater extent. The most potent AMO inhibitors contain hairpin structural elements that flank the AMO. At present, it is not yet clear why and how the flanking regions enhance AMO activities. One possible interpretation of this finding is that secondary structure enhances miRNA–RISC interactions with AMOs.

The procedures for designing such extended AMOs are described below.

1. Select a miRNA of interest;
2. Design an oligonucleotide fragment exactly complementary (base-pairing or matching) to the guide strand of the mature miRNA;
3. Add 5′- and 3′-flanking sequences to the above-designed fragment according to the following notes (Fig. 7.1).
   (a) The flanking sequence at both sides should be equal in length;
   (b) The optimal length of the flanking sequence is 16 nts;
   (c) The flanking sequence can be complementary to regions adjacent to the mature miRNA sequence in the target pre-miRNA or can be arbitrary;
   (d) Oligonucleotides with polypyrimidine stretches are generally considered to be less structured than sequences having an equal distribution of purines and pyrimidines;
   (e) Equal hairpin structures in the flanking sequences enhance the stability and functionality of the antisense.
4. Chemically synthesize the designed oligoribonucleotide fragment with 2′-O-methyl modification; this is your AMO. You may attach a C7 amino linker to the 3′-end of AMO if you plan to introduce post-synthetic conjugation of non-nucleotidic residues such as biotin for monitoring transfection;
5. For negative control, synthesize an oligonucleotide fragment complementary to the passenger strand of the miRNA;
6. Transfection of AMOs (10–50 nM) in cultured cells or in tissues by in vivo gene transfer techniques. If using cationic lipid, we generally cannot transfect more than 100 nM AMS of any chemistry without non-specific toxicity.

Horwich and Zamore (2008) summarized that an effective AMO is (a) resistant to nonspecific cellular ribonucleases, (b) resistant to miRNA-directed cleavage by RISC and (c) binds miRNAs in RISCs with high affinity, effectively out-competing binding to target mRNAs.

7.2.2 Modifying AMOs

Modification to stabilize AMOs to nuclease degradation and improve affinity for target miRNAs are necessary for their miRNA-antagonizing activities in cell culture and animals. In addition to 2′-*O*-methyl (2′-OMe) modification as mentioned above, other 2′-sugar modifications, including 2′-*O*-methoxyethyl (2′-MOE), 2′-flouro (2′-F) and locked nucleic acid (LNA), as well as phosphorothioate backbone modification, have also been examined (Leaman et al. 2005; Davis et al. 2006; Meister et al. 2004; Hutvágner et al. 2004; Ørom et al. 2006; Krutzfeldt et al. 2005; Kloosterman et al. 2007).

1. 2′-O-methyl (2′-OMe) modification. The first AMOs reported to inhibit miRNA activity in vitro were 2′-OMe modified and these are commercially available

from a variety of sources. 2′-O-Methyl modification is frequently used to protect oligoribonucleotides from degradation in cell extracts or cultured cells because of the resistance to cleavage by both RISC and other cellular ribonucleases (reviewed in Lamond and Sproat 1993; Verma and Eckstein 1998; Hutvágner et al. 2004; Meister et al. 2004; Inoue et al. 1987). 2′-O-Methyl oligoribonucleotides are also well-known for their rapid and very stable hybridization to single-stranded RNA (Cummins et al. 1995; Majlessi et al. 1998). Moreover, 2′-O-methyl-modified RNA–RNA hybrids are more thermodynamically stable than either RNA–RNA or DNA–RNA duplexes (Inoue et al. 1987; Tsourkas et al. 2002).

2. Other 2′ modifications. Other 2′ modifications, such as 2′-O-methoxyethyl AMOs and AMOs incorporating pyrimidines bearing 2′-O-fluoro (Esau et al. 2006; Davis et al. 2006), all have been shown to improve affinity to the targeted miRNAs to some degrees, despite that their anti-miRNA activities have not correlated perfectly with affinity (Davis et al. 2006), suggesting there are other variables to the interaction important for effective inhibition.
3. Phosphorothioate backbone. The phosphorothioate backbone is known to be the primary determinant of oligonucleotide distribution in vivo and can be delivered to the liver, kidney, bone marrow and adipose tissue following parenteral administration (Geary et al. 2001). Phosphorothioate modified antisense, while reducing affinity to target RNA somewhat, confers significant stability to nuclease degradation and may be necessary for long-term assays in cell culture. In the absence of formulation, the phosphorothioate backbone modification is essential for in vivo delivery of AMOs to tissues, as the phosphorothioate promotes protein binding and delays plasma clearance (Geary et al. 2001). Esau et al. (2006) demonstrated inhibition of miR-122, an abundant liver-specific miRNA, implicated in fatty acid and cholesterol metabolism as well as hepatitis C viral replication, in normal mice using a 2′-MOe modified AMO with a full phosphorothioate backbone. The AMO was delivered intraperitoneally, twice weekly in saline. After four weeks of treatment at doses as low as 12.5 mg/kg, increased levels of miR-122 target gene mRNAs and reduced plasma cholesterol levels were also observed, as well as a reduction in miR-122 levels.
4. Locked Nucleic Acid (LNA) modification. LNA modified AMOs, which are commercially available, also showed significantly better activity than the 2′-OMe (Elmén et al. 2005). LNA is defined as oligonucleotides containing one or more LNA monomers, the 2′-O, 4′-C-methylene-β-D-ribofuranosyl nucleosides. A major structural characteristic of LNA is the close resemblance to the natural nucleic acids because methylene group-linked O2′ and the C4′ atoms introduce a conformational lock of the molecule into a near perfect N-type conformation adopted by RNA (Saenger 1984; Jepsen and Wengel 2004). This modification tremendously enhances affinity to the target miRNA. A fully modified LNA sequence has been reported to be fully resistant to the 3′-exonuclease (Frieden et al. 2003). End-blocked sequence, i.e., LNA-DNA-LNA gapmers display a high stability in human serum compared to similar 2′-OMe modified sequences (Kurreck et al. 2002). Recently, Elmén et al. (2008) reported that a systemically

administered 16-nt, unconjugated LNA-AMO towards miR-122 leads to specific, dose-dependent silencing of miR-122 in the liver and shows no hepatotoxicity in mice. Moreover, the efficacy of LNA/OMe eximer in anti-miRNA activity has also been tested (Fabani and Gait 2008).

5. Cholesterol conjugation. When attached with cholesterol moieties to permit easier entry into the cell, an AMO is called 'antagomiRs'. AntagomiRs can be used to bind with target miRNAs and have been shown to be capable of in vivo silencing expression of miRNAs in the mouse liver and heart (Krutzfeldt et al. 2005). Conjugation of a cholesterol to the 3′end of a 2′-OMe modified AMO with two or three phosphorothioate modifications on each end has been reported to facilitate in vivo delivery of the AMO targeting miR-122 into the liver (Krutzfeldt et al. 2005). After three days of daily intravenous (i.v.) administration of 80 mg kg^{-1} AMO to normal mice, increased levels of miR-122 target gene mRNAs were observed in the liver, as well as reduced plasma cholesterol and an apparent degradation of the miRNA. A more recent study claimed that 3′-cholesterol-modified AMOs have enhanced potency, allowing miRNA inhibition for at least 7 d from a single transfection (Horwich and Zamore 2008).

However, in a study reported by Vermeulen et al. (2007), AMOs containing an assortment of chemical modifications (e.g., 2′-O-methyl, 2′-fluoro, phosphorothioate, or a combination of these modifications) failed to identify any pattern of modifications that significantly enhanced function and in some cases, identified modifications that greatly reduced the ability to inhibit RISC function.

7.2.3 Monitoring Delivery Efficiency of AMOs

A crucial step for effective inhibition of miRNAs activity in cultured cells is optimization of AMO transfection. For most adherent cell types, a commercially available cationic lipid (lipofectamine 2000 from Invitrogen) is usually an effective delivery method (Dass et al. 2002; Dass et al. 2002; Zimmermann et al. 2006). Optimization of transfection should be done for each cell type, as there are a wide range of optimal transfection reagents or delivery methods for different cell types. For cells that are refractory to lipid transfection, electroporation is usually a good alternative. Again, this will require optimization for each cell type.

Evaluating transfection efficiency can be done by monitoring reduction of a miRNA by an AMO or by measuring uptake of a fluorescently labeled AMO. A phenotypic readout is not a reliable endpoint for optimizing transfection. We used the following methods to monitor AMO transfection efficiency and subcellular distribution (Gao et al. 2006; Yang et al. 2007).

7.2.3.1 Measurement of Uptake of Fluorescent AMOs

1. When designing an AMO, attach a C7 amino linker to the 3′-end of AMO. FITC label the AMO. FITC is a small fluorescein isothiocyanate molecule and is typically conjugated to proteins via primary amines (i.e., lysines);

2. One day before treatment, plate cells in 24-well format with 1×10^5 cells/well in 500 μl;
3. On the day of treatment, incubate the cells with FITC-labeled AMO at a desired concentration (i.e., 100 nM), in the presence of Lipofectamine 2000 for 4 h;
4. Following incubation, harvest the cells with PBS-EDTA and wash them twice with PBS and then soak them in TBS+50 mM glycine for 10 min;
5. Determine the amount of internalized AMO by flow cytometry at selected time points after transfection;
6. At last, the percentage of cells with successful uptake of FITC-labeled AMO can be determined by counting cells with clear yellow staining. The values are plotted as a function of time after transfection to obtain an AMO uptake curve. The concentration of AMO associated with the cells can be estimated by interpolation from a standard curve of known FITC (Molecular Probes, Eugene, OR).

7.2.3.2 Subcellular Localization of Transfected AMOs

1. When designing an AMO, attach a C7 amino linker to the 3′-end of AMO;
2. Label AMOs with Alexa Fluor 488 using *ULYSIS* Nucleic Acid Labeling kits (Invitrogen);
3. Purify the labeled AMOs with Micro Bio-Spin 30 columns (Bio-Rad Laboratories);
4. Transfect the cells grown on sterile coverslips in 12-well plates with the AMOs. At the selected time points after transfection, wash the cells twice with phosphate-buffered saline (PBS) and fix them with 2% paraformaldehyde for 20 min;
5. To visualize nuclear DNA, equilibrate the fixed cells in 2x SSC solution (0.3 M NaCl and 0.03 M sodium citrate, pH 7.0) and incubate with 100 g ml^{-1} DNase-free RNase in 2x SSC for 20 min at 37°C;
6. Rinse the sample three times in 2xSSC and incubate with 5 μM propidium iodide for 30 min at RT;
7. Mount the coverslips onto slides with DABCO medium. Examine the samples under a laser scanning confocal microscope with Alexa Fluor 488 (excitation at 492 nm and emission at 520 nm) or with PI (excitation at 535 nm and emission at 617 nm).

7.2.4 Evaluating Functional Effectiveness of AMOs

Measuring functional outcomes produced by AMOs, such as developmental events, cell proliferation, apoptosis and other phenotypic readouts as appropriate, is a key to determine the functional effectiveness of AMOs. However, experimental validation of AMOs must also be done at different levels to ensure non-toxicity and specificity of actions.

1. Knockdown of targeted miRNAs. Quantifying the level of the targeted miRNAs by Northern blotting analysis or real-time RT-PCR methods. The level or detectability is expected to be reduced by at least 95% to be considered effective;
2. Upregulation of protein expression by AMO. Monitoring changes of expression at the protein level of the target gene of the targeted miRNAs by Western blotting analysis or other protein quantification assays such as enzyme activity assays. The protein level is expected to increase by at least 30% to be considered effective;
3. Heterologous reporter assay. Measuring the efficacy of AMOs using a luciferase reporter (or other reporter genes) bearing miRNA target sequences cloned into its 3′UTR. An AMO is expected to inhibit the repression effect of its endogenous target miRNA on the reporter activities or in other words, to enhance the reporter activities by reliving repression by its endogenous target miRNA. It is also necessary to determine if an AMO is able to antagonize the repressive effect on luciferase activities produced by an exogenously applied target miRNA through cotransfection process;
4. Immobilized 2′-O-methyl oligonucleotide capture of RISC. If RISC activity needs to be assessed, the procedures described by Hutvágner et al. (2004), Elbashir et al. (2001) and Nykänen et al. (2001) are useful.

    (a) Incubate biotinylated AMO (10 pM) for 1 h on ice in lysis buffer containing 2 mM DTT with 50 μl of Dynabeads M280 (as a suspension as provided by the manufacturer; Dynal, Oslo, Norway) to immobilize the oligonucleotide on the beads. To ensure that the AMO remained in excess when more than 50 nM target miRNA was used, 20 pM of biotinylated AMO was immobilized;
    (b) For RISC capture assays, pre-incubate miRNA in a standard 50 μl in vitro RNAi reaction for 15 min at 25°C;
    (c) Add the immobilized AMO to the reaction and incubate for another 1 h;
    (d) Collect beads after incubation using a magnetic stand (Dynal). Recover the unbound supernatant and an aliquot assayed for RISC activity to confirm that RISC depletion is complete;
    (e) Wash the beads three times with ice-cold lysis buffer containing 0.1% (w/v) NP-40 and 2 mM DTT, followed by a wash without NP-40. To determine the amount of RISC formed, input and bound radioactivity is determined by scintillation counting (Beckman Instruments);

5. Negative control experiments. It is essential to have a negative control AMO of the same chemistry, as different chemical modifications may have different thresholds for toxicity.

7.3 Principle of Actions

AMOs are artificially designed 2′-O-methyl oligonucleotides fully complementary to their target miRNAs. Once introduced into cells, AMOs bind efficiently by base-pairing with the guide strand of their target miRNAs in miRISC. In this way, they mediate potent, irreversible, miRNA-specific inhibition of miRNA function, providing a powerful "loss-of-function" strategy for miRNA interference. Typically, AMOs are transiently transfected into cells, correspondingly providing a transient derepression of miRNA targets. An AMO inhibits the function of its targeted miRNA through multiple mechanisms.

1. Breaking down the targeted miRNAs. The most fundamental and widely observed mode of action of AMOs is to cause degradation of their targeted miRNAs with unknown mechanisms (Esau et al. 2006; Krutzfeldt et al. 2005, 2007). Several observations indicate that the phenotypic outcomes resulting from antisense injection are rescued by genomic overexpression of the cognate miRNAs.
2. Sequestering target miRNAs. By binding to the guide strand of the target miRNA, an AMO can effectively prevent the binding of its targeted miRNA from binding to the target gene, as a competitive inhibitor to block the function of the targeted miRNA. This is in pretty much the same way as the blocking actions on transcription factors by decoy oligodeoxynucleotides (Gao et al. 2006; Lin et al. 2007). It has indeed been observed that with some modified AMOs, potent inhibition of miRNA activity can be elicited without a decrease in miRNA levels in vivo, suggesting that AMOs can sequester the miRNA without causing degradation (Chan et al. 2005; Esau et al. 2006). Therefore, measuring miRNA levels is sometimes not a reliable measure of miRNA inhibition.
3. Disrupting RISC-mediated targeting in the cytoplasm. AMOs can act as non-cleavable substrates (antisense) of RISC by replacing the passenger strand of the targeted mature miRNA to disrupt the RISC-mediated targeting (Meister et al. 2004; Hutvágner et al. 2004).
4. Disrupting Drosha processing in the nucleus. Treatment of whole cells with fluorescently labeled molecules shows a predominant nuclear localization pattern. This observation suggests that AMOs may be able to act by inhibiting miRNA function in the nucleus by disrupting Drosha processing, as an alternative to other mechanisms.

7.4 Applications

AMOs have become a necessary and most commonly used tool for miRNA research and have been proposed to be potential therapeutic agents as well. Since the invention of this technology in 2004, no less than 90% of published experimental studies have used AMOs in one form or the other as a 'miRNA-loss-of-function' approach to delineate miRNA targets and function or to rescue miRNA-repressed gene expression and reverse pathological processes.

Modified AMOs are currently the most readily available tools for miRNA inhibition and several groups have used different backbone modifications of these antisense oligos to successfully inhibit miRNAs in cell culture. Two independent groups have also shown successful miRNA inhibition in the liver following systemic and intraperitoneal delivery. In addition, a recent study showed that 2′-O-methyl-modified AMOs flanked by hairpin sequences give enhanced inhibition of miRNAs in cell culture (Vermeulen et al. 2007). Esau et al. (2006) focused on the miRNA pathway itself as a potential target for pharmacological intervention in an animal model. They injected a miR-122 AMO (an antisense oligonucleotide with 2′-O-methoxyethyl phosphororothiate) into mice that resulted in decreased plasma cholesterol levels and hepatic fatty acid and cholesterol synthesis. Nevertheless, the same limitations encountered with the application of synthetic miRNA duplexes are encountered in the applications of antagomiRs, namely their effective delivery into target tissues.

1. To validate the miRNA targets. Understanding the biological function of miRNAs requires knowledge of their mRNA targets. To validate the theoretically predicted miRNA targets, new and rapid methods for sequence-specific inactivation of miRNAs are needed. The ability of an AMO to block the binding of its targeted miRNA to a gene helps determining whether this gene is a target of that miRNA. The method thus provides a means of connecting miRNA target identification with biological function and of distinguishing phenocritical from neutral targets.
2. To validate miRNA function. The AMO technique can be used not only for genome-wide phenotypic screening but also for a detailed subsequent characterization of phenotypes using markers and epistasis experiments, thereby placing miRNAs more precisely within biological processes and pathways. Use of AMOs can aid to define the cause-effect relationship between a miRNA and a biological event, thereby the cellular function of miRNAs.

Cheng et al. (2005) synthesized over 90 miRNA inhibitors and used these to screen for miRNAs that were involved in cell growth and apoptosis processes that are among the widely studied gene cell pathways, which have direct relevance to cancer and development. From these experiments, they identified a complexity of activity for miRNAs in two different types of cancer-derived cell lines, HeLa and A549 cells. In HeLa, they found that inhibition of miR-95, 124, 125, 133, 134, 144, 150, 152, 187, 190, 191, 192, 193, 204, 211, 218, 220, 296 and 299 caused a decrease in cell growth and that inhibition of miR-21 and miR-24 had a profound increase in cell growth. On the other hand, miR-7, 19a, 23, 24, 134, 140, 150, 192 and 193 downregulated cell-growth and miR-107, 132, 155, 181, 191, 194, 203, 215 and 301 increased cell growth when inhibited in A549 cells. Common miRNAs that decreased cell growth include miR-134, 192 and 193. They also used screening to identify miRNAs involved in induction or inhibition of steady state levels of apoptosis in HeLa cells. For this study, the authors identified miR-1d, 7, 148, 204, 210, 218, 296 and 381 as hits that increased the level of apoptosis and miR-214 that decreased apoptosis. In another case, miR-218 caused a decrease in cell growth in HeLa but its inhibition increased the level of apoptosis, suggesting that inhibition of miR-218 may be inhibiting cell growth by inducing apoptosis.

3. To achieve upregulation of the cognate target protein. The AMO technology is unique in its action and outcome of action. AMOs act on miRNAs in a

'loss-of-function' manner to silence these miRNAs but they result in 'gain-of-function' of target protein-coding genes via relieving the post-transcriptional repressive effects of their targeted miRNAs on these protein-coding genes. In this way, AMOs upregulate gene expression.

This property of AMOs can be utilized in many situations for the benefits of treating human disease in cases when upregulation of gene expression is desirable. One most obvious application of the AMO approach is to upregulate expression of tumor suppressors or apoptotic molecules that are repressed by miRNAs in cancer cells. It has been demonstrated that knockdown of miR-21 in cultured glioblastoma cells and in MCF-7 human breast cancer cells resulted in a significant drop in cell number. This reduction was accompanied by increases in caspase-3 and -7 enzymatic activities and TUNNEL staining (Chan et al. 2005; Corsten et al. 2007). miR-21 knockdown also has been shown to decrease expression of antiapoptotic Bcl-2 protein (Si et al. 2007). Knockdown of the muscle-specific miRNA miR-1 by its specific AMO has also been reported to enhance expression of connexin-43 and improve cardiac conduction so as to limit arrhythmogenesis in infarct heart (Yang et al. 2007).

'Gain-of-function' of target protein-coding genes produced by AMOs can also be used to validate the cellular function and target gene of miRNAs. If an AMO upregulates tumor suppressor gene, then it is very likely that the counterpart of this AMO, the targeted miRNA, acts to repress the expression of this tumor suppressor gene.

4. To reverse the pathological process. Many miRNAs have been implicated in human disease or in animal models of disease, such as cancer, cardiovascular disturbances (e.g., ischemic arrhythmogenesis, cardiac hypertrophy and heart failure), metabolic disorders (e.g., type 2 diabetes), viral diseases, Alzheimer's disease, etc. Targeting pertinent miRNAs with AMOs has been shown to be effective in reversing these pathological processes. For example, we have demonstrated the ability of AMOs to antagonize the ischemic arrhythmias induced by miR-1 (Yang et al. 2007), to abolish the abnormal QT prolongation caused by miR-133 in diabetic hearts (Xiao et al. 2007a) and to reverse miR-1-induced heart rate-reducing effects (Xiao et al. 2007b).

7.5 Advantages and Limitations

The AMO approach as a miRNAi technology or targeting-miRNA technology has many advantages over other available gene-interference strategies, such as the conventional antisense (ASO) technique, siRNA technique, decoy ODN technique, etc. It shows stoichiometric and reliable inhibition of the targeted miRNA and can thus be applied to studies of miRNA functions and validation of putative target genes.

1. Easy to design and produce. An AMO is simply a 2′-OMe oligonucleotides fragment fully antisense to the guide strand of target miRNA. There is no need to search for the gene-specific sequences for actions like conventional ASO and

siRNAs and there are no concerns about the secondary structures of the fragment. A few companies provide service for synthesis of AMO with one fragment for as little as $20.

2. Reasonable specificity of targeting miRNA. An AMO is entirely complementary to its target miRNA and thus has a highest binding affinity to this miRNA than to any other sequence. AMOs have been used to inhibit exogenously introduced miRNAs and endogenous miRNAs. The AMO approach is sufficiently specific to distinguish between more distantly related miRNAs, as in the case of miRNA families: family members show distinct phenotypes and target interaction profiles.
3. High efficiency of targeting miRNA. We and others have consistently observed virtually knockout degree of suppression of miRNA expression by AMOs in cell lines; an AMO is able to achieve nearly 100% knockdown of its targeted miRNA, as determined by quantitative real-time RT-PCR (Xiao et al. 2007a, b; Yang et al. 2007; Matsubara et al. 2007).
4. Upregulation of gene expression. An interesting and important point about the AMO approach is that unlike other forms of gene interference (siRNA, conventional antisense, etc.) that are used to silence gene expression, the action of AMOs is to silence miRNAs but to upregulate gene expression by relieving the repressive effect of miRNAs on their target protein-coding genes. In some cases, upregulation of gene expression is desirable. For instance, enhancing expression of tumor suppressor genes is one of the strategies for cancer therapy.
5. Knockout versus knockdown. Difficulty of generating miRNA knockout animals due to multicopy nature of many miRNAs. For example, the miR-1 family is comprised of the miR-1 subfamily and miR-206, with the former consisting of 2 transcripts, miR-1-1 and miR-1–2 that possess an identical mature sequence but are encoded by distinct genes located on chromosomes 2 and 18, respectively. The miR-133 family is comprised of miR-133a-1, miR-133a-2 and miR-133b, with miR-133a expressed from bicistronic units together with the miR-1 subfamily and miR-133b together with miR-206 (Chen et al. 2006; Rao et al. 2006). The resulting mature products from the miR-133 family are either identical or have only 1-nt difference.

The obvious alternative to inhibiting miRNA function by AMOs is to generate genomic mutants. However, reverse genetic analysis on a large-scale is laborious and particularly challenging in the case of many miRNAs, due to redundancy and complex genomic organization (e.g., clustering of unrelated miRNAs and dispersal of multicopy miRNAs among different regions of the genome) (Lai 2003). While the generation of genomic knockouts remains the biological gold standard desirable for an in-depth study of individual miRNAs, its technical difficulty paired with the complex organization of miRNA genes makes it a high-risk, time-consuming approach. Application of AMOs can entirely inhibit the function of targeted miRNAs no matter how many genomic copies as AMOs act on mature miRNAs. I believe that AMO-mediated depletion provides a powerful means for the investigation of miRNA function. At present, AMO perhaps is the first choice of 'loss-of-function' approach in miRNA research. Compared with gene knockout approaches, miRNA-knockdown also has the advantage of being simple, quick and low cost and can be used for rapid functional studies of individual miRNAs.

It should always be cognizant that the AMO technology has some innate limitations, which must be vigilantly considered when using it.

1. Unlike conventional antisense acting on protein-coding genes, AMOs are designed against miRNAs. As the effects of miRNAs are not gene-specific, the effects of AMOs cannot be either. An important concept is that the action of AMOs is miRNA-specific but not specific towards a particular protein-coding gene; by knocking down a miRNA, an AMO is deemed to affect all target genes of its targeted miRNA. Under certain circumstances when actions directed to a particular gene are desirable, AMOs may not be the choice.
2. A potential limitation of a hybridization-based approach is cross-reactivity. While an AMO may be sufficiently specific to distinguish between more distantly related miRNAs, it might cross hybridize to miRNAs with very close sequence similarity. In this way, an AMO might cause some non-specific unwanted effects.
3. An AMO directs to a specific miRNA. Unfortunately, a large number of miRNA exist in a cell and many of them share similar sequences. The miRNAs with similar sequences, particularly in their seed sites, are likely to have very similar target interactions and may in fact have to be removed as a group to reveal their function. With respect to the mode of action, AMOs are incapable of simultaneously targeting a group of miRNAs. Providentially, other miRNAi technologies related to AMO have been developed to compensate this weakness, including the Multiple-Target AMO technology (MT-AMO technology) and MiRNA Sponge technology, as described in Chaps. 7 and 8, respectively.

From a therapeutic standpoint, however, manipulating the levels of one miRNA may have only a limited effect on the expression of the desired target gene.

References

Boutla A, Delidakis C, Tabler M (2003) Developmental defects by antisense mediated inactivation of micro-RNAs 2 and 13 in Drosophila and the identification of putative target genes. Nucleic Acids Res 31:4973–4980.

Chan JA, Krichevsky AM, Kosik KS (2005) MicroRNA-21 is an antiapoptotic factor in human glioblastoma cells. Cancer Res 65:6029–6033.

Chen JF, Mandel EM, Thomson JM, Wu Q, Callis TE, Hammond SM, Conlon FL, Wang DZ (2006) The role of microRNA-1 and microRNA-133 in skeletal muscle proliferation and differentiation. Nat Genet 38:228–233.

Chen JF, Murchison EP, Tang R, Callis TE, Tatsuguchi M, Deng Z, Rojas M, Hammond SM, Schneider MD, Selzman CH, Meissner G, Patterson C, Hannon GJ, Wang DZ (2008) Targeted deletion of Dicer in the heart leads to dilated cardiomyopathy and heart failure. Proc Natl Acad Sci USA 105:2111–2116.

Cheng AM, Byrom MW, Shelton J, Ford LP (2005) Antisense inhibition of human miRNAs and indications for an involvement of miRNA in cell growth and apoptosis. Nucleic Acids Res 33:1290–1297.

Corsten MF, Miranda R, Kasmieh R, Krichevsky AM, Weissleder R, Shah K (2007) MicroRNA-21 knockdown disrupts glioma growth in vivo and displays synergistic cytotoxicity with neural precursor cell delivered S-TRAIL in human gliomas. Cancer Res 67:8994–9000.

Cummins LL, Owens SR, Risen LM, Lesnik EA, Freier SM, McGee D, Guinosso CJ, Cook PD (1995) Characterization of fully 2_-modified oligoribonucleotide hetero- and homoduplex hybridization and nuclease sensitivity. Nucleic Acids Res 23:2019–2024.

Dass CR (2002) Liposome-mediated delivery of oligodeoxynucleotides in vivo. Drug Deliv 9:169–180.

Dass CR, Walker TL, Burton MA (2002). Liposomes containing cationic dimethyl dioctadecyl ammonium bromide: formulation, quality control, and lipofection efficiency. Drug Deliv 9:11–18.

Davis S, Lollo B, Freier S, Esau C (2006) Improved targeting of miRNA with antisense oligonucleotides. Nucleic Acids Res 34:2294–2304.

Elbashir SM, Lendeckel W, Tuschl T (2001) RNA interference is mediated by 21- and 22-nucleotide RNAs. Genes Dev 15:188–200.

Elmén J, Lindow M, Silahtaroglu A, Bak M, Christensen M, Lind-Thomsen A, Hedtjärn M, Hansen JB, Hansen HF, Straarup EM, McCullagh K, Kearney P, Kauppinen S (2008) Antagonism of microRNA-122 in mice by systemically administered LNA-antimiR leads to up-regulation of a large set of predicted target mRNAs in the liver. Nucleic Acids Res 36:1153–1162.

Elmén J, Thonberg H, Ljungberg K, Frieden M, Westergaard M, Xu Y, Wahren B, Liang Z, Ørum H, Koch T, Wahlestedt C (2005) Locked nucleic acid (LNA) mediated improvements in siRNA stability and functionality. Nucleic Acids Res 33:439–447.

Esau CC (2008) Inhibition of microRNA with antisense oligonucleotides. Methods 44:55–60.

Esau C, Davis S, Murray SF, Yu XX, Pandey SK, Pear M, Watts L, Booten SL, Graham M, McKay R, Subramaniam A, Propp S, Lollo BA, Freier S, Bennett CF, Bhanot S, Monia BP (2006) miR-122 regulation of lipid metabolism revealed by in vivo antisense targeting. Cell Metab 3:87–98.

Fabani MM, Gait MJ (2008) miR-122 targeting with LNA/2′-O methyl oligonucleotide mixmers, peptide nucleic acids (PNA), and PNA-peptide conjugates. RNA 14:336–346.

Frieden M, Christensen SM, Mikkelsen ND, Rosenbohm C, Thrue CA, Westergaard M, Hansen HF, Ørum H, Koch T (2003) Expanding the design horizon of antisense oligonucleotides with alpha-L-LNA. Nucleic Acids Res 31:6365–6372.

Geary RS, Watanabe TA, Truong L, Freier S, Lesnik EA, Sioufi NB, Sasmor H, Manoharan M, Levin AA (2001) Pharmacokinetic properties of 2′-O-(2-methoxyethyl)-modified oligonucleotide analogs in rats. J Pharmacol Exp Ther 296:890–897.

Gao H, Xiao J, Yang B, Sun Q, Lin H, Bai Y, Yang L, Wang H, Wang Z (2006) A single decoy oligodeoxynucleotides targeting multiple oncoproteins produces strong anti-cancer effects. Mol Pharmacol 70:1621–1629.

Hammond SM (2006) MicroRNA therapeutics: a new niche for antisense nucleic acids. TiMM 12:99–101.

Horwich MD, Zamore PD (2008) Design and delivery of antisense oligonucleotides to block microRNA function in cultured Drosophila and human cells. Nat Protoc 3:1537–1549.

Hutvágner G, Simard MJ, Mello CC, Zamore PD (2004) Sequence-Specific Inhibition of Small RNA Function. PLoS Biol 2:465–475.

Inoue H. Hayase Y, Imura A, Iwai S, Miura K, Ohtsuka E (1987) Synthesis and hybridization studies on two complementary nona(2′-O-methyl)ribonucleotides. Nucleic Acids Res 15:6131–6148.

Jepsen JS, Wengel L (2004) LNA-antisense rivals siRNA for gene silencing. Curr Opin Drug Discovery Dev 7:188–194.

Kloosterman WP, Lagendijk AK, Ketting RF, Moulton JD, Plasterk RH (2007) Targeted inhibition of miRNA maturation with morpholinos reveals a role for miR-375 in pancreatic islet development. PLoS Biol 5:e203.

Krutzfeldt J, Kuwajima S, Braich R, Rajeev KG, Pena J, Tuschl T, Manoharan M, Stoffel M (2007) Specificity, duplex degradation and subcellular localization of antagomirs. Nucleic Acids Res 35:2885–2892.

Krutzfeldt J, Rajewsky N, Braich R, Rajeev KG, Tuschl T, Manoharan M, Stoffel M (2005) Silencing of microRNAs in vivo with 'antagomirs'. Nature 438:685–689.

Kurreck J, Wyszko E, Gillen C, Erdmann VA (2002) Design of antisense oligonucleotides stabilized by locked nucleic acids. Nucleic Acids Res 30:1911–1198.

Lai EC (2003) microRNAs: runts of the genome assert themselves. Curr Biol 13:R925–R936.

Lamond AI, Sproat BS (1993) Antisense oligonucleotides made of 2′-O-alkyl RNA: Their properties and applications in RNA biochemistry. FEBS Lett 325:123–127.

Leaman D, Chen PY, Fak J, Yalcin A, Pearce M, Unnerstall U, Marks DS, Sander C, Tuschl T, Gaul U (2005) Antisense-mediated depletion reveals essential and specific functions of microRNAs in Drosophila development. Cell 121:1097–1108.

Lee Y, Ahn C, Han J, Choi H, Kim J, Yim J, Lee J, Provost P, Rådmark O, Kim S, Kim VN (2003) The nuclear RNase III Drosha initiates microRNA processing. Nature 425:415–419.

Lee Y, Hur I, Park SY, Kim YK, Suh MR, Kim VN (2006) The role of PACT in the RNA silencing pathway. EMBO J 25:522–532.

Lin H, Xiao J, Luo X, Xu C, Gao H, Wang H, Yang B, Wang Z (2007) Overexpression HERG K^+channel gene mediates cell-growth signals on activation of oncoproteins Sp1 and NF-κB and inactivation of tumor suppressor Nkx3.1. J Cell Physiol 212:137–147.

Majlessi M, Nelson NC, Becker MM (1998) Advantages of 2′-O-methyl oligoribonucleotide probes for detecting RNA targets. Nucleic Acids Res 26:2224–2229.

Matsubara H, Takeuchi T, Nishikawa E, Yanagisawa K, Hayashita Y, Ebi H, Yamada H, Suzuki M, Nagino M, Nimura Y, Osada H, Takahashi T (2007) Apoptosis induction by antisense oligonucleotides against miR-17-5p and miR-20a in lung cancers overexpressing miR-17-92. Oncogene 26:6099–6105.

Meister G, Landthaler M, Dorsett Y, Tuschl T (2004) Sequence-specific inhibition of microRNA- and siRNA-induced RNA silencing. RNA 10:544–550.

Murchison EP, Partridge JF, Tam OH, Cheloufi S, Hannon GJ (2005) Characterization of Dicer-deficient murine embryonic stem cells. Proc Natl Acad Sci USA 102:12135–12140.

Nykänen A, Haley B, Zamore PD (2001) ATP requirements and small interfering RNA structure in the RNA interference pathway. Cell 107:309–321.

Ørom UA, Kauppinen S, Lund AH (2006) LNA-modified oligonucleotides mediate specific inhibition of microRNA function. Gene 372:137–141.

Rao PK, Kumar RM, Farkhondeh M, Baskerville S, Lodish HF (2006) Myogenic factors that regulate expression of muscle-specific microRNAs. Proc Natl Acad Sci USA 103:8721–8726.

Saenger W (1984) Principles of nucleic acid structure. Springer, New York.

Schaefer A, O'Carroll D, Tan CL, Hillman D, Sugimori M, Llinas R, Greengard P (2007) Cerebellar neurodegeneration in the absence of microRNAs. J Exp Med 204:1553–1558.

Si ML, Zhu S, Wu H, Lu Z, Wu F, Mo YY (2007) miR-21-mediated tumor growth. Oncogene 26:2799–1803.

Tsourkas A, Behlke MA, Bao G (2002) Hybridization of 2′-O-methyl and 2′-deoxy molecular beacons to RNA and DNA targets. Nucleic Acids Res 30: 5168–5174.

Verma S, Eckstein F (1998) Modified oligonucleotides: Synthesis and strategy for users. Annu Rev Biochem 67:99–134.

Vermeulen A, Robertson B, Dalby AB, Marshall WS, Karpilow J, Leake D, Khvorova A, Baskerville S (2007) Double-stranded regions are essential design components of potent inhibitors of RISC function. RNA 13:723–730.

Wang Z, Luo X, Lu Y, Yang B (2008) miRNAs at the heart of the matter. J Mol Med 86:772–783.

Weiler J, Hunziker J, Hall J (2006) Anti-miRNA oligonucleotides (AMOs): ammunition to target miRNAs implicated in human disease? Gene Ther 13:496–502.

Wu H, Lima WF, Zhang H, Fan A, Sun H, Crooke ST (2004) Determination of the role of the human RNase H1 in the pharmacology of DNA-like antisense drugs. J Biol Chem 279:17181–17189.

Xiao J, Luo X, Lin H, Xu C, Gao H, Wang H, Yang B, Wang Z (2007a) MicroRNA miR-133 represses HERG K^+channel expression contributing to QT prolongation in diabetic hearts. J Biol Chem 282:12363–12367.

Xiao J, Yang B, Lin H, Lu Y, Luo X, Wang Z (2007b) Novel approaches for gene-specific interference via manipulating actions of microRNAs: examination on the pacemaker channel genes HCN2 and HCN4. J Cell Physiol 212:285–292.

Yang B, Lin H, Xiao J, Luo X, Li B, Lu Y, Wang H, Wang Z (2007) The muscle-specific microRNA miR-1 causes cardiac arrhythmias by targeting GJA1 and KCNJ2 genes. Nat Med 13:486–491.

Zimmermann TS, Lee AC, Akinc A, Bramlage B, Bumcrot D, Fedoruk MN, Harborth J, Heyes JA, Jeffs LB, John M, Judge AD, Lam K, McClintock K, Nechev LV, Palmer LR, Racie T, Röhl I, Seiffert S, Shanmugam S, Sood V, Soutschek J, Toudjarska I, Wheat AJ, Yaworski E, Zedalis W, Koteliansky V, Manoharan M, Vornlocher HP, MacLachlan I (2006) RNAi-mediated gene silencing in non-human primates. Nature 441:111–114.

Chapter 8
Multiple-Target Anti-miRNA Antisense Oligonucleotides Technology

Abstract The multiple-target AMO technology or MT-AMO technology is an innovative strategy that confers a single AMO fragment the capability of targeting multiple miRNAs. These modified AMOs are single-stranded 2′-*O*-methyl-modified oligoribonucleotides carrying multiple AMO units that are engineered into a single unit and are able to simultaneously silence multiple target miRNAs or multiple miRNA seed families. Studies suggest the MT-AMO is an improved approach for miRNA target finding and miRNA function validation; it not only enhances the effectiveness of targeting miRNAs but also confers diversity of actions. It has been successfully used to identify target genes and cellular function of several oncogenic miRNAs and of the muscle-specific miRNAs [Lu Y, Xiao J, Lin H, Bai Y, Luo X, Wang Z, Yang B, Nucleic Acid Res. 2009]. This novel strategy may find its broad application as a useful tool in miRNA research for exploring biological processes involving multiple miRNAs and multiple genes and potential as a miRNA therapy for human disease such as cancer and cardiac disorders. This technology was developed by my research laboratory in collaboration with Yang's group (Lu Y, Xiao J, Lin H, Bai Y, Luo X, Wang Z, Yang B 2009). The MT-AMO technology belongs to the "targeting-miRNA" and "miRNA-loss-of-function" strategy. The MT-AMO technology is based on the 'One-Drug, Multiple-Target' concept (see Sect. 2.1.3 for detail).

8.1 Introduction

Anti-miRNA antisense inhibitors (AMOs) have demonstrated their utility in miRNA research and potential in miRNA therapy. However, it has become clear that a particular condition may be associated with multiple miRNAs and a given gene may be regulated by multiple miRNAs. For example, a study directed to the human heart identified 67 significantly upregulated miRNAs and 43 significantly downregulated miRNAs in failing left ventricles vs. normal hearts (Thum et al. 2007). No less than five different miRNAs have been shown to be critically involved in cardiac hypertrophy. Similarly, Volinia et al. (2006) conducted a large-scale miRNA analysis

Z. Wang, *MicroRNA Interference Technologies*,
DOI: 10.1007/978-3-642-00489-6_8,

on 540 solid tumor samples including lung, breast, stomach, prostate, colon and pancreatic tumors. Their survey revealed 15 miRNAs upregulated and 12 downregulated in human breast cancer tissues. Similar changes of multiple miRNAs were also found in five other solid tumor types. Many of these miRNAs have been reported to regulate cell proliferation or apoptosis and some of them have been considered oncogenic miRNAs or tumor suppressor miRNAs. These facts raise two questions. Is targeting a single miRNA sufficient for tackling a particular pathological condition? Does simultaneously targeting multiple miRNAs relevant to a particular condition offer an improved outcome than targeting a single miRNA using the regular AMO techniques?

This observation has led many in the field to speculate that miRNAs work in a combinatorial fashion and act in concert to target a single transcript (Krek et al. 2005). Moreover, many miRNAs are members of families that share a seed sequence but may have one or more nucleotide changes in the remaining sequence. These related miRNAs are expected to regulate similar target mRNAs and if co-expressed in a tissue it may be necessary to inhibit all of them at once to observe phenotypic effects. There are few examples where overexpression or knockdown of a single miRNA results in a measurable phenotype (Esau et al. 2004; Poy et al. 2004; Yekta et al. 2004; Johnson et al. 2005; Lee et al. 2005; Lim et al. 2005; Chen et al. 2006; Schratt et al. 2006; Yang et al. 2007).

These properties of miRNA regulation may well create some uncertainties of outcomes by applying the AMO technology to silence miRNAs since knocking down a single miRNA may not be sufficient to achieve the expected interference of cellular process and gene expression, which are regulated by multiple miRNAs. Tools that allow multi-miRNA knockdown will be essential for identification and validation of miRNA targets. For such applications, cotransfection of multiple AMOs targeting various isoforms is possible (Bommer et al. 2007). In this respect, genetic approaches are superior for studying individual miRNA family members, whereas miRNA sponges (introduced in the next chapter) or multiple AMOs are appropriate for studying miRNA families whose members only contain a common seed sequence; single AMO studies may simplify the study of nearly identical miRNA paralogs. Combinations of AMOs targeting unrelated miRNAs have also been used to disrupt more than one miRNA in the same transfected cells, obviating the need to make and combine multiple genetic knockouts. Because combinatorial control of targets by miRNAs may be common (Bartel and Chen 2004), this approach may prove particularly important for uncovering networks of miRNAs that act together. Vermeulen et al. (2007) showed that cotransfection of a six-AMO mixture can effectively derepress reporters for each individual miRNA. Functional studies also suggest that cotransfection of AMOs is effective. For instance, Pedersen et al. (2007) tested the efficacy of five interferon-β induced miRNAs with seed matches to Hepatitis C genes in antiviral activity; indeed, simultaneous cotransfection of all five AMOs but not controls, significantly enhanced Hepatitis C RNA production.

Co-application of multiple AMOs while effective in some cases may be problematic in that control of equal transfection efficiency is difficult, if not impossible. To tackle the problem, an innovative strategy, the multiple-target AMO technology or MT-AMO technology, which confers a single AMO fragment the capability of targeting multiple miRNAs has been developed in our laboratories (Lu et al. 2009). This modified AMO carries multiple antisense units that are engineered into a single unit, which is able to simultaneously silence multiple target miRNAs. Studies suggest the MT-AMO is an improved approach for miRNA target-gene finding and for studying the function of miRNAs. This novel strategy may find its broad application as a useful tool in miRNA research, for exploring biological processes involving multiple miRNAs and multiple genes and potential as a miRNA therapy for human disease such as cancer and cardiac disorders.

8.2 Protocols

8.2.1 Designing MT-AMOs

1. Select a particular gene, a particular cellular function or a particular disease for your study. Then determine, based on published studies, the miRNAs that can target the gene under test or that are known to be associated with or implicated in this gene or cellular function or disease;
2. Design anti-miRNA oligonucleotides fragments (AMOs) exactly antisense to the selected miRNAs for your study. Then link these AMOs in the orientation of $5'$-end to $3'$-end together to form a long MT-AMO. Note that an AMO can be a RNA or a DNA (ODN, oligodeoxynucleotide); for commercial synthesis, an ODN is less costly than a RNA fragment;
3. A linker (CTTAAATG) may be inserted in-between every two AMOs. This design gives you a MT-AMO that contains all the AMOs needed to knockdown the miRNAs relevant to the gene or cellular function or disease of your interest;
4. Synthesize the designed MT-AMO fragment with chemical modifications as described in Chap. 7;
5. Store the MT-AMO construct at -80°C for future use;
6. Construct a negative control MT-AMO (NC MT-AMO) for verifying the efficacy and specificity of the effects of the MT-AMO. This NC MT-AMO should be designed based on the sequence of the MT-AMO; simply modifying the MT-AMO to contain ~5-nts mismatches at the $5'$-end "seed site" to make it expectedly able to destruct the binding to the target miRNAs (Fig. 8.1).

8.2.2 Validating MT-AMOs

Readers are referred to the steps described in Sect. 7.2.4 for validating the effectiveness of MT-AMOs.

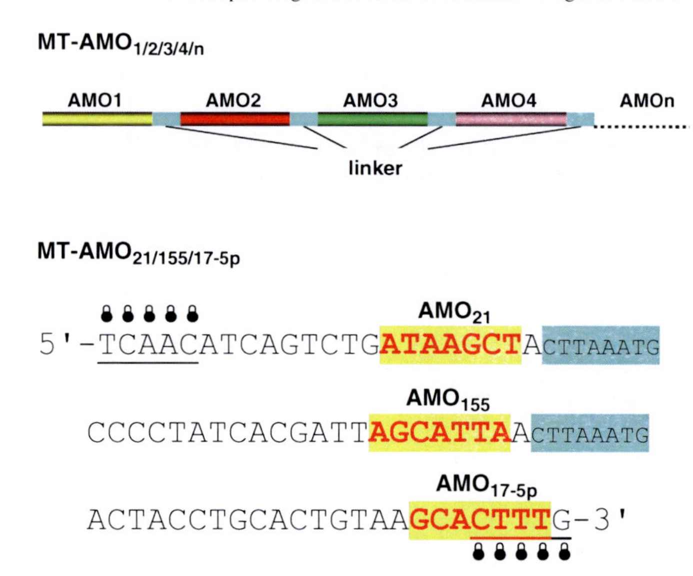

Fig. 8.1 Design of Multiple-Target Anti-miRNA Antisense Oligomers (MT-AMO). *Upper*: Schematic illustration of a MT-AMO. A MT-AMO incorporates multiple AMO units targeting different miRNAs. The number of AMO units in a MT-AMO is in theory unlimited; but longer sequence of a MT-AMO may create unfavorable secondary structure and may also increase the difficulty of being uptaken into a cell. *Lower*: Example of a MT-AMO designed to carry 3 AMOs targeting miR-21, miR-155 and miR-17-5p used in our previous study (Lu Y, Xiao J, Lin H, Bai Y, Luo X, Wang Z, Yang B 2009). The sequences corresponding to the seed sites of the miRNAs are in red boldface and are highlighted in yellow. To enhance the stability and affinity, MT-AMO is chemically modified to have 5 nts at both ends locked with methylene bridges (LNA). An eight-nucleotide linker (highlighted by blue) is inserted to connect the two adjacent AMO units

8.3 Principle of Actions

Most human diseases are multifactorial and multistep processes. Targeting a single factor (molecule) may not be adequate and certainly not optimal in disease therapy, because single agents are limited by incomplete efficacy and dose-limiting adverse effects. If related factors are concomitantly attacked, better outcomes are expected and the current combination pharmacotherapy was developed for this reason: a combination of two or more drugs or therapeutic agents given as a single treatment that successfully saves lives. The "drug cocktail" therapy of AIDS is one example of such a strategy (Henkel 1999) and similar approaches have been used for a variety of other diseases, including cancers (Konlee 1998; Charpentier 2002; Ogihara 2003; Kumar 2005; Nabholtz and Gligorov 2005). However, the current drug-cocktail therapy is costly and may involve complicated treatment regimen, undesired drug-drug interactions and increased side effects (Konlee 1998). There is a need to develop a strategy to avoid these problems and a "one-drug, multiple-target" strategy is highly desirable. However, it is nearly impossible to confer to single compounds the ability to act on multiple target molecules with the traditional pharmaceutical approaches or the currently known antigene strategies.

The MT-AMO technology was developed to enable an AMO to target multiple miRNAs in order to effectively interfere with expression of a protein-coding gene that is regulated by these multiple miRNAs so as to effectively alter a relevant cellular function and physiological process. The basic mechanism of action of MT-AMOs is identical to the single AMOs described in Chap. 7.

8.4 Applications

In addition to the applications of regular AMOs stated in Chap. 7, our MT-AMO technology, a 'one-drug, multiple-target' strategy, bears further superiority. In theory, it mimics the well-known drug cocktail therapy. But it is devoid of the weaknesses of the drug cocktail therapy, involving complicated treatment regimens, undesirable drug-drug interactions and increased side effects. The MT-AMO technology offers resourceful combinations of varying AMOs for concomitantly targeting multiple miRNAs for studying or treating biological and pathophysiological processes involving multiple factors.

We have validated the technique with two separate MT-AMOs: anti-miR-21/anit-miR-155/anti-miR-17-5p ($\text{MT-AMO}_{21/155/17}$) and anti-miR-1/anti-miR-133 ($\text{MT-AMO}_{1/133}$) (Lu et al. 2009).

miR-21, miR-155 and miR-17-5p are proven oncogenic miRNAs overexpressing in several solid cancers (Iorio et al. 2005; Volinia et al. 2006) and miR-1 and miR-133 are muscle-specific miRNAs crucial for myogenesis. We first tested the ability of $\text{MT-AMO}_{21/155/17}$ and $\text{MT-AMO}_{1/133}$ to inactivate their respective target miRNAs by luciferase reporter assays. We then evaluated the effects of the two MT-AMOs on the protein levels of predicted target genes tumor suppressor genes TGFBI, APC and BCL2L11 for miR-21, miR-155 and miR-17-5p, respectively and HCN2 (a subunit of pacemaker channel) and Cav1.2 (α-subunit of L-type Ca^{2+} channel encoded by CACNA1C) for miR-1 and miR-133, respectively. We further demonstrated that $\text{MT-AMO}_{21/155/17}$ induced MCF-7 human breast cancer cell death with a greater efficacy and potency than the regular singular AMOs (Lu et al. 2009).

8.5 Advantages and Limitations

The MT-AMO technology is a simple and efficient approach to study a complicated, multifactorial cellular process that is regulated by multiple miRNAs. When containing AMO units towards miRNA seed families, MT-AMOs are capable of knocking down members of multiple miRNA seed families.

However, many important issues remain unresolved in validating the MT-AMO technology as a gene therapy strategy. The optimal combination of targets for an MT-AMO remains unknown. In the example study described above, we tested "three-in-one" MT-AMOs. In theory, "*N*-in-one" MT-AMOs (N could be any number of AMO units) can be designed to include more relevant target TFs; however, larger MT-AMOs may hinder their penetration into the cells and compromise the effectiveness. More rigorous studies are warranted to define the optimal combination of length and accessibility of MT-AMOs to optimize desired effectiveness. Our previous work did not allow us to draw any conclusions as to what the optimal organization is for multiple AMO units to be placed in a single MT-AMO molecule. Moreover, efficient delivery of MT-AMOs into a cell is another challenge to using MT-AMOs as therapeutic agents, as in other nucleotide-based technologies such as siRNA, antisense, ribozyme, aptamers, etc. Another difficulty is to maintain an effective concentration of MT-AMO within a cell for a sufficient period of time. At present, investigation on modifications of MT-AMOs to enhance efficiency of transfection and to strengthen the stability within a cell so as to prolong the duration of actions is an active field of research. Constructing MT-AMO into virus vectors, such as adenovirus, lentivirus, etc., might be a reasonable approach to offset the weakness of the nucleotide technologies.

References

Bartel DP, Chen CZ (2004) Micromanagers of gene expression: The potentially widespread influence of metazoan microRNAs. Nat Rev Genet 5:396–400.

Bommer GT, Gerin I, Feng Y, Kaczorowski AJ, Kuick R, Love RE, Zhai Y, Giordano TJ, Qin ZS, Moore BB, MacDougald OA, Cho KR, Fearon ER (2007) p53-mediated activation of miRNA34 candidate tumor-suppressor genes. Curr Biol 17:1298–1307.

Charpentier G (2002) Oral combination therapy for type 2 diabetes. Diabetes Metab Res Rev 18(Suppl 3):S70–S76.

Chen JF, Mandel EM, Thomson JM, Wu Q, Callis TE, Hammond SM, Conlon FL, Wang DZ (2006) The role of microRNA-1 and microRNA-133 in skeletal muscle proliferation and differentiation. Nat Genet 38:228–233.

Esau C, Kang X, Peralta E, Hanson E, Marcusson EG, Ravichandran LV, Sun Y, Koo S, Perera RJ, Jain R, Dean NM, Freier SM, Bennett CF, Lollo B, Griffey R (2004) MicroRNA-143 regulates adipocyte differentiation. J Biol Chem 279:52361–52365.

Iorio MV, Ferracin M, Liu CG, Veronese A, Spizzo R, Sabbioni S, Magri E, Pedriali M, Fabbri M, Campiglio M, Menard S, Palazzo JP, Rosenberg A, Musiani P, Volinia S, Nenci I, Calin GA, Querzoli P, Negrini M, Croce CM (2005) MicroRNA gene expression deregulation in human breast cancer. Cancer Res 65:7065–7070.

Johnson SM, Grosshans H, Shingara J, Byrom M, Jarvis R, Cheng A, Labourier E, Reinert KL, Brown D, Slack FJ (2005) RAS is regulated by the let-7 microRNA family. Cell 120:635–647.

Henkel J (1999) Attacking AIDS with a 'cocktail' therapy? FDA Consum 33:12–17.

Konlee M (1998) An evaluation of drug cocktail combinations for their immunological value in preventing/remitting opportunistic infections. Posit Health News 16:2–4.

Krek A, Grun D, Poy M, Wolf R, Rosenberg L, Epstein E, MacMenamin P, da Piedade I, Gunsalus K, Stoffel M, Rajewsky N (2005) Combinatorial microRNA target predictions. Nat Genet 37:495–500.

Kumar P (2005) Combination treatment significantly enhances the efficacy of antitumor therapy by preferentially targeting angiogenesis. Lab Investig 85:756–767.

Lee YS, Kim HK, Chung S, Kim KS, Dutta A (2005) Depletion of human micro-RNA miR-125b reveals that it is critical for the proliferation of differentiated cells but not for the down-regulation of putative targets during differentiation. J Biol Chem 280:16635–16641.

Lim LP, Lau NC, Garrett-Engele P, Grimson A, Schelter JM, Castle J, Bartel DP, Linsley PS, Johnson JM (2005) Microarray analysis shows that some microRNAs downregulate large numbers of target mRNAs. Nature 433:769–773.

Lu Y, Xiao J, Lin H, Bai Y, Luo X, Wang Z, Yang B (2009) Complex antisense inhibitors offer a superior approach for microRNA research and therapy. Nucleic Acids Res 37:e24–e33.

Nabholtz JM, Gligorov J (2005) Docetaxel/trastuzumab combination therapy for the treatment of breast cancer. Expert Opin Pharmacother 6:1555–1564.

Ogihara T (2003) The combination therapy of hypertension to prevent cardiovascular events (COPE) trial: Rationale and design. Hypertens Res 28:331–338.

Pedersen IM, Cheng G, Wieland S, Volinia S, Croce CM, Chisari FV, David M (2007) Interferon modulation of cellular microRNAs as an antiviral mechanism. Nature 449:919–922.

Poy MN, Eliasson L, Krutzfeldt J, Kuwajima S, Ma X, Macdonald PE, Pfeffer S, Tuschl T, Rajewsky N, Rorsman P, Stoffel M (2004) A pancreatic islet-specific microRNA regulates insulin secretion. Nature 432:226–230.

Schratt GM, Tuebing F, Nigh EA, Kane CG, Sabatini ME, Kiebler M, Greenberg ME (2006) A brain-specific microRNA regulates dendritic spine development. Nature 439:283–289.

Thum T, Galuppo P, Wolf C, Fiedler J, Kneitz S, van Laake LW, Doevendans PA, Mummery CL, Borlak J, Haverich A, Gross C, Engelhardt S, Ertl G, Bauersachs J (2007) MicroRNAs in the human heart: a clue to fetal gene reprogramming in heart failure. Circulation 116:258–267.

Vermeulen A, Robertson B, Dalby AB, Marshall WS, Karpilow J, Leake D, Khvorova A, Baskerville S (2007) Double-stranded regions are essential design components of potent inhibitors of RISC function. RNA 13:723–730.

Volinia S, Calin GA, Liu CG, Ambs S, Cimmino A, Petrocca F, Visone R, Iorio M, Roldo C, Ferracin M, Prueitt RL, Yanaihara N, Lanza G, Scarpa A, Vecchione A, Negrini M, Harris CC, Croce CM (2006) A microRNA expression signature of human solid tumors defines cancer gene targets. Proc Natl Acad Sci USA 103:2257–2261.

Yang B, Lin H, Xiao J, Luo X, Li B, Lu Y, Wang H, Wang Z (2007) The muscle-specific microRNA miR-1 causes cardiac arrhythmias by targeting GJA1 and KCNJ2 genes. Nat Med 13:486–491.

Yekta S, Shih IH, Bartel DP (2004) MicroRNA-directed cleavage of HOXB8 mRNA. Science 304:594–596.

Chapter 9
miRNA Sponge Technology

Abstract MiRNA Sponge technology is an innovative approach used to generate RNAs containing multiple, tandem binding sites for a miRNA seed family of interest and able to target all members of that miRNA seed family. When vectors encoding the miRNA sponges are transiently transfected into cultured cells, they depress miRNA targets as strongly as the conventional AMOs described in Chap. 6. The major advancement of this technique over the AMO technique is that it can better inhibit functional classes of miRNAs than do AMOs that are designed to block single miRNA sequences. The main principle of the miRNA Sponge technology is identical to the MT-AMO technology described in Chap. 7: targeting multiple miRNAs. The miRNA Sponge technology was established by Sharp's laboratory in 2007 [Ebert MS, Neilson JR, Sharp PA, Nat Methods 4:721–726 2007; Hammond SM Nat Methods 4:694–695, 2007]. Similar to the AMO approach, miRNA Sponge technology belongs to the "targeting-miRNA" and "miRNA-loss-of-function" strategy. The miRNA Sponge technology complies with the 'Single-Drug, Multiple-Target' [Gao H, Xiao J, Sun Q, Lin H, Bai Y, Yang L, Yang B, Wang H, Wang Z, Mol Pharmacol, 70:1621–1629, 2006] and 'miRNA Seed Family' concepts (see Sect. 2.1.4 for detail).

9.1 Introduction

As reasoned in Sect. 2.1.4 and Chap. 8, many miRNAs are members of families that share a seed sequence but may have one or more nucleotide changes in the remaining sequence (Bommer et al. 2007; Pedersen et al. 2007; Vermeulen et al. 2007). Moreover, many miRNAs are expressed from multiple genomic loci (Wang 2008). To achieve adequate miRNA-loss-of-function for elucidating a certain cellular process, the conventional AMO strategy falls short in dealing with multiple miRNAs. On the other hand, creating genetic knockouts to determine the function of miRNA families is difficult, as individual miRNAs expressed from multiple genomic loci or from multiple members of a same miRNA seed family may repress a common set

Z. Wang, *MicroRNA Interference Technologies*,
DOI: 10.1007/978-3-642-00489-6_9, © Springer-Verlag Berlin Heidelberg 2009

of targets containing a complementary seed sequence. Thus, a method for inhibiting these functional classes of paralogous miRNAs in vivo is needed.

For this reason, Ebert et al. (2007) invented an innovative anti-miRNA approach termed ‘miRNA sponges’. The idea behind it is to produce a single specie of RNAs containing multiple, tandem binding sites for a miRNA seed family of interest, in order to target all members of that miRNA seed family, taking advantage of the fact that the interaction between miRNA and target is nucleated by and largely dependent on base-pairing in the seed region (positions 2–8 of the miRNA). The authors constructed sponges by inserting tandemly arrayed miRNA binding sites into the 3′UTR of a reporter gene encoding destabilized GFP driven by the CMV promoter, which can yield abundant expression of the competitive inhibitor transcripts.

9.2 Protocols

The miRNA Sponge technology shares many similarities with the AMO and the MT-AMO technologies in that they are all antisense to miRNAs acting on their target miRNA by base-pairing mechanisms and producing miRNA-loss-of-function effects. There is therefore no surprise that the protocols involved in the generation and application of the miRNA inhibitors with these distinct technologies are more or less the same. The procedures described in this chapter are primarily based upon the studies reported by Ebert et al. (2007).

9.2.1 Designing miRNA Sponges

1. Select a particular miRNA seed family of your interest for study and further select a member from this seed family for designing a miRNA sponge. For example, miR-17-5p, miR-20 and miR-17–92 are within the same seed family and miR-30c, miR-30d and miR-30e are within another seed family. Pick any one member from a seed family as a template for designing a miRNA sponge;
2. Determine a binding site for the selected miRNA that is a sequence perfectly complementary to the selected member of the miRNA seed family, just like designing an AMO described in Chap. 7. But this AMO or binding site carries mismatches in the middle portion to create a bulge (4–7 A:G/G:A wobble pairs) at positions 9–12 when binding to its target miRNAs. Each of such an AMO is considered one unit that can be recognized as a binding site by multiple miRNAs that share the seed;

(The bulge is for preventing RNA interference–type cleavage and degradation of the sponge RNA through endonucleolytic cleavage by Argonaute 2. For both sponge classes, sponges with 4–7 bulged binding sites will produce stronger derepressive effects than sponges with two perfect binding sites. This difference may be due to the availability of more binding sites in the bulged sponges and/or to the greater

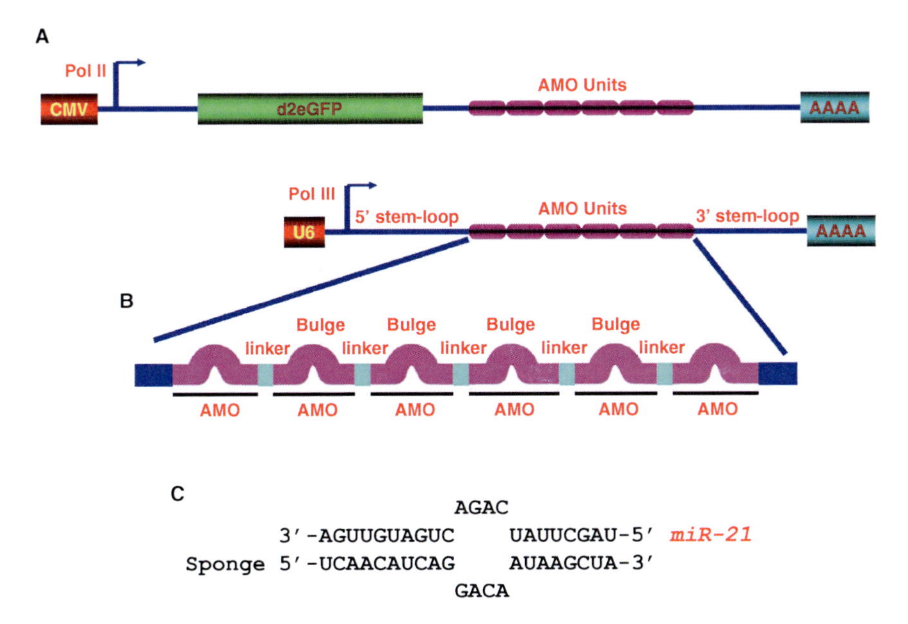

Fig. 9.1 Design of miRNA sponges according to the protocols by Ebert et al. (2007). (**a**) Construction of GFP sponges by inserting multiple identical miRNA binding sites (AMO units) into the 3′UTR of a 2-h destabilized GFP (d2eGFP) reporter gene driven by the CMV promoter. (**b**) Construction of U6 sponges by subcloning multiple identical AMO units into a vector containing a U6 snRNA promoter with 5′ and 3′ stem-loop elements. (**c**) The imperfect pairing between a miRNA and a sponge with bulged binding sites. An example is shown for miR-21. The bulge is designed to protect against endonucleolytic cleavage by Argonaute 2

stability expected of bulged sponge RNAs compared to sponge RNAs that can be cleaved by miRNA–loaded Argonaute 2. The imperfect pairing between a miRNA and a sponge with bulged binding sites is diagrammed for miR-21 in Fig. 9.1);

3. Connect the AMO units in tandem to construct multiple miRNA binding sites in a single fragment. The 5′-end AMO unit and the 3′-end AMO unit should each carry restriction enzyme recognition sequence for cloning (e.g., 5′-XhoI and 3′-ApaI in the study by Ebert et al. (2007)).
4. Ebert et al. (2007) examined 7 miR-20 binding sites in their study. Addition of more miRNA binding sites to the sponge might increase the potency of the inhibitor. These authors also tested miRNA sponges with 6, 10 and 18 sites. However, they found only a marginal increase in activity above 6 sites, with apparently saturating effect. A spacer with an irrelevant sequence might be inserted in-between binding sites to optimize the binding of miRNPs to every possible binding site. But previous results suggest that nearby sites are fully functional (Doench and Sharp 2004);
5. Design a complementary strand. Synthesize both strands. Anneal the two strands to form a double-stranded construct;

6. Construct a miRNA sponge containing RNA polymerase II promoter plasmid (pcDNA5-CMV-d2eGFP vector, Invitrogen) by inserting the fragment into the 3′UTR (at XhoI and ApaI cloning site) of a reporter gene encoding destabilized GFP driven by the CMV promoter.
7. Alternatively, one can construct a Pol III plasmid (RNA polymerase III promoter) carrying miRNA sponges to take advantage of strong Pol III promoters, which are known to drive expression of the most-abundant cellular RNAs. Subclone the fragment carrying tandemly arrayed miRNA binding sites from the GFP sponge construct described in (**3**) into a modified U6 small nuclear RNA promoter-terminator vector, which produces short (<300 nt) RNAs with stem-loop elements (Paul et al. 2003). As construct lacks an open reading frame, the U6 sponges become substrates for miRNA binding but not for translation or translational repression;
8. As a negative control, construct a sponge with repeated binding sites complementary to an artificial miRNA based on a sequence from genes but not complementary to any known miRNA;
9. Perform functional assays to validate the miRNA sponges, as outlined below.

9.2.2 Validating miRNA Sponges

1. Transfect the sponge plasmids into selected cell lines. Ideally, the cells for transfection should contain abundant endogenous target miRNAs, otherwise target synthetic miRNA (SC-miRNA; see Chap. 3) must be cotransfected. Measure the levels of targeted miRNAs using real-time qRT-PCR methods or Northern blot. The levels of miRNA members within the targeted miRNA seed family are expected to be knocked down as strongly as the conventional AMOs;
2. Construct a miRNA binding site–carrying luciferase reporter gene vector by inserting the binding site of the target miRNA seed family into the multiple cloning sites downstream the luciferase gene (3′UTR region). For example, you can insert a 22-nt fragment containing the 5′ portion 1–8 nts of the miR-17-5p seed family and an arbitrary sequence of 16 nts. Then, cotransfect a sponge plasmid and the luciferase vector at a ratio of 8:1, sponge plasmid to target plasmid, in a selected cell line. 24∼48 h after transfection, measure luciferase activity. The luciferase activity is expected to increase with application of the miRNA sponge but not with cotransfection of negative control sponges;
3. The GFP gene in the plasmid can be used for monitoring the transfection efficiency of the sponge plasmids and for tracking those cells that express high levels of the inhibitor RNA. Simply conduct quantification of GFP transcripts by real-time PCR in relation to GFP plasmid standards. It was estimated that GFP mRNAs in transiently transfected 293T cells were at least 1,000–2,000 per cell. Then, this level of expression of a miRNA sponge containing seven binding sites targeting a miRNA seed family allows inhibition of approximately

10^4 miRNAs per cell, which would be sufficient to inhibit most miRNAs in most cell types;
4. Test the ability of sponges to derepress natural miRNA target genes using Western blot and qRT-PCR methods. The target genes of the selected miRNA seed family are expected to be upregulated in their expression at the protein level. The transcript level of miRNA target genes may or may not be altered depending upon the overall complementarity between the sponge sequence and the target mRNA sequence.

9.3 Principle of Actions

miRNA sponges are transcripts expressed from strong promoters, containing multiple, tandem binding sites to a selected member of a miRNA seed family of interest. miRNA sponges act by mechanisms similar to AMO and MT-AMO; they can sequester targeted miRNAs, disrupt miRISC-mediated targeting in the cytoplasm and disrupt Drosha processing in the nucleus. But unlike AMO and MT-AMO, miRNA sponges with bulges are in theory not able to induce degradation of their targeted miRNA. Detailed comparisons among the miRNA Sponge, AMO and MT-AMO technologies are given in Table 9.1.

9.4 Applications

Like the AMO and the MT-AMO approaches, the miRNA Sponge technology is able to interfere with the function of natural, endogenous target miRNAs, the target genes of the miRNAs and therefore can be used for target validation and phenotypic analysis.

Table 9.1 Comparisons of major characteristics among the miRNA Sponge, AMO and MT-AMO technologies

	miRNA sponge	MT-AMO	AMO
Composition	mRNA containing a 3′UTR	DNA or RNA oligos	RNA or DNA oligo
Number of binding sites	Multiple (homogenous)	Multiple (homogenous or heterogeneous)	Single
Complementarity	Seed site (8 nts) complementarity	Full (22 nts) complementarity	Seed Full (22 nts) complementarity
Targeted miRNA	Targeted miRNAs intact	Targeted miRNAs degradation	Targeted miRNA degradation
Specificity	miRNA-specific & miRNA seed family-specific	miRNA-specific & miRNA seed family- specific	miRNA-specific
Outcome	Derepression of proteins	Derepression of proteins	Derepression of proteins

A potential extension of the miRNA Sponge technology would be to express sponges from stably integrated transgenes in vivo. This can be applied to studying long-term effects of miRNA-loss-of-function in cell lines, drug-inducible miRNA sponges in xenograft models to investigate miRNA contributions to tumorigenesis and to treat cancer; bone marrow reconstitution approaches to investigate miRNAs roles in immune cell development; and maybe germline transgenic sponge mice to ascertain the functions of miRNAs families at cell, tissue, organ and organism levels.

9.5 Advantages and Problems

1. miRNA sponges are at least as effective as the AMO technology in antagonizing their target miRNAs;
2. As an alternative to AMOs, miRNA sponges can be expressed in cells, as RNAs produced from transgenes; these competitive inhibitors are transcripts expressed from strong promoters, containing multiple, tandem binding sites to a miRNA of interest;
3. They specifically inhibit miRNAs with a complementary heptameric seed, such that a single sponge can be used to block an entire miRNA seed family;
4. miRNA sponges can be made stably expressed in cell lines from multicopy chromosomal insertions (by cotransfecting 293T cells with linearized GFP sponge plasmids and a puromycin selection marker) Ebert et al. (2007) reported that the stable *miR*-16 sponge–expressing cell line allowed threefold higher expression of a *miR*-16 target, relative to cells transiently transfected with sponge plasmids.

In terms of the limitation of the approach, the same concept that is applied to AMOs and MT-AMOs is also applicable to miRNA sponges; that is, the action of miRNA sponges is miRNA-specific but not specific towards a particular protein-coding gene; by knocking down a miRNA or a miRNA seed family, a miRNA sponge is deemed to affect all target genes of its targeted miRNAs. In many situations, this action generates undesirable effects. In such a case, a different approach, MiRNA-Masking Antisense Oligonucleotides (miR-Mask) technology (Chap. 10) can be employed instead.

References

Bommer GT, Gerin I, Feng Y, Kaczorowski AJ, Kuick R, Love RE, Zhai Y, Giordano TJ, Qin ZS, Moore BB, MacDougald OA, Cho KR, Fearon ER (2007) p53-mediated activation of miRNA34 candidate tumor-suppressor genes. Curr Biol 17:1298–1307.

Doench JG, Sharp PA (2004) Specificity of microRNA target selection in translational repression. Genes Dev 18:504–511.

Ebert MS, Neilson JR, Sharp PA (2007) MicroRNA sponges: Competitive inhibitors of small RNAs in mammalian cells. Nat Methods 4:721–726.

Gao H, Xiao J, Sun Q, Lin H, Bai Y, Yang L, Yang B, Wang H, Wang Z (2006) A single decoy oligodeoxynucleotides targeting multiple oncoproteins produces strong anticancer effects. Mol Pharmacol 70:1621–1629.

Hammond SM (2007) Soaking up small RNAs. Nat Methods 4:694–695.

Paul CP, Good PD, Li SX, Kleihauer A, Rossi JJ, Engelke DR (2003) Localized expression of small RNA inhibitors in human cells. Mol Ther 7:237–247.

Pedersen IM, Cheng G, Wieland S, Volinia S, Croce CM, Chisari FV, David M (2007) Interferon modulation of cellular microRNAs as an antiviral mechanism. Nature 449:919–922.

Vermeulen A, Robertson B, Dalby AB, Marshall WS, Karpilow J, Leake D, Khvorova A, Baskerville S (2007) Double-stranded regions are essential design components of potent inhibitors of RISC function. RNA 13:723–730.

Chapter 10
miRNA-Masking Antisense Oligonucleotides Technology

Abstract miRNA-Masking Antisense Oligonucleotides Technology (miR-Mask) is an AMO approach of a different sort. A standard miR-Mask is a single-stranded 2′-*O*-methyl-modified oligoribonucleotide (or other chemically modified), which is a 22-nt antisense to a protein-coding mRNA as a target for an endogenous miRNA of interest. Instead of binding to the target miRNA like an AMO, a miR-Mask does not directly interact with its target miRNA but binds to the binding site of that miRNA in the 3′UTR of the target mRNA by a fully complementary mechanism. In this way, the miR-Mask covers up the access of its target miRNA to the binding site to derepress its target gene (mRNA) via blocking the action of its target miRNA. The anti-miRNA action of a miR-Mask is gene-specific because it is designed to be fully complementary to the target mRNA sequence of a miRNA. The anti-miRNA action of a miR-Mask is also miRNA-specific as well because it is designed to target the binding site of that particular miRNA. The miR-Mask approach is a valuable supplement to the AMO technique; while AMO is indispensable for studying the overall function of a miRNA, the miR-Mask might be more appropriate for studying the specific outcome of regulation of the target gene by the miRNA. This technology was first established by my research group in 2007 [Xiao J, Yang B, Lin H, Lu Y, Luo X, Wang Z, J Cell Physiol 212:285–292, 2007b] and a similar approach with the same concept was subsequently reported by Schier's laboratory [Choi WY, Giraldez AJ, Schier AF, Science 318:271–274 2007]. Similar to the AMO approach, miR-Mask technology belongs to the "targeting-miRNA" and "miRNA-loss-of-function" strategy.

10.1 Introduction

Each single miRNA may regulate as many as 1,000 protein-coding genes and each gene may be regulated by multiple miRNAs. This implies that the action of miRNAs is binding sequence-specific but not gene-specific; similarly, the action of AMO, thereby MT-AMO and miRNA Sponge, is miRNA-specific but not gene-specific

Z. Wang, *MicroRNA Interference Technologies*,
DOI: 10.1007/978-3-642-00489-6_10, © Springer-Verlag Berlin Heidelberg 2009

either. These properties of miRNAs and AMOs may present obstacles for development as therapeutic agents, since they may elicit unwanted side effects and toxicity through their non-gene-specific functional profiles. For example, the muscle-specific miRNA miR-1 has the potential to post-transcriptionally repress a number of ion channel genes including cardiac sodium channel gene SCNCA5, pacemaker channel gene HCN4, gap junction channel connexin 43, inward rectifier K^+channel KCNJ2 and voltage-dependent K^+channel KCND2. Based on this targeting, miR-1 is expected to affect cardiac electrophysiology. Taking the concept of "miRNA as a Regulator of a Cellular Function", one can just focus on what miR-1 does on the cardiac electrophysiology. However, if one wants to understand the mechanisms using miR-1 loss-of-function strategy, the AMO, MT-AMO and miR-Sponge technologies will all fall short due to their lack of gene specificity. By knocking down miR-1, one will potentially alter the expression of all miR-1 target genes mentioned above.

To emasculate the problem, we have developed the miRNA-Masking Antisense Oligonucleotides Technology (miR-Mask), which provides a gene-specific strategy for studying miRNA function and mechanisms. Using the miR-Mask approach, one is now able to dissect the role of each of the target genes of a miRNA, say the ion channel genes for miR-1. For instance, one can use a miR-Mask on SCNCA5 to explore the role of miR-1 regulation of this sodium channel on cardiac electrophysiology. Soon after our publication on this technology, Choi et al. (2007) published a study using essentially the same strategy and they named the technology "Target Protector". For convenience and clarity, I suggest using miR-Mask as a unified name.

10.2 Protocols

1. Analyze sequences to identify the binding site(s) for an endogenous miRNA of interest, in the 3′UTR of the target mRNA. For example, we have shown that the muscle-specific miRNA miR-133 represses the protein expression of HERG K^+channel gene KCNH2, contributing to the increased risk of pathologic long QT syndrome in diabetic cardiomyopathy (Xiao et al. 2007a). We performed a study to see if we could relieve the repression to reduce the arrhythmogenic potential in diabetic hearts. For this end, the first step is to analyze the 3′UTR of KCNH2 around the region containing the binding sequence for miR-133;
2. Design an oligonucleotide fragment of around 22 nts or longer, fully antisense to the region covering the binding sequence of the miRNA of interest;
3. Blast search to verify the uniqueness of the fragment to ensure the gene-specificity. Once confirmed, the fragment is considered a miR-Mask;
4. Chemical synthesize the miR-Mask using services provided by commercial companies such as IDT Technologies or Ambion. Remember to chemical modify the oligonucleotide as described in Chap. 6;

5. Transfect the miR-Mask into cells to study the enhancing effects on protein expression of the target gene. In the case of a miR-Mask for KCNH2 and miR-133, we expect to see an increase in the HERG protein level.

10.3 Principle of Actions

The miR-Mask strategy was developed to interfere with function of the endogenous miRNAs in a gene-specific and miRNA-specific manner. The idea is to use a miR-Mask to regulate the protein expression of the target gene (mRNA) by interfering with the action of a particular miRNA on this gene. Two prerequisites must be fulfilled: (1) the presence of a recognition motif for a miRNA within the 3′UTR of the target gene and (2) the presence of a fragment with unique sequences containing the miRNA-biding motif, which is sufficiently long (~22 nts) for miR-Mask binding. The first prerequisite ensures the miRNA specificity of miR-Mask action and the second the gene specificity of miR-Mask action.

A miRNA-mask is designed to fully base-pair with the binding motif of an endogenous miRNA in the 3′UTR of the target mRNA. Upon delivery into the cell, the miR-Mask is expected to bind itself to the region to block the access of that endogenous miRNA to the site of action. In this way, the miR-Mask disrupts miRNA:mRNA interaction to relieve the repressive action of the miRNA on the target gene to promote the protein expression of that gene.

The miR-Mask approach defers from the AMO approach in several aspects, despite that they both can result in enhancement of gene expression by removing the repressive effects of a particular miRNA on protein translation of the target mRNA (Table 10.1).

Table 10.1 Comparisons among miR-Mask, AMO, and conventional antisense oligomer (ASO)

	miR-Mask	AMO	ASO
Structure	RNA or DNA	RNA or DNA	DNA
Targeting	mRNA (3′UTR)	miRNA	mRNA (CdR)
Specificity	mRNA-specific (Gene-specific)	miRNA-specific (Non-gene-specific)	mRNA-specific (Gene-specific)
Mode of action	miRNA intact mRNA intact	Cleaving target miRNA	Cleaving target mRNA
Mechanism	Masking miRNA binding site in target mRNA	Relieving translational repression	Blocking translational process
Outcome	Protein expression ⬆ (mRNA level no change)	Protein expression ⬇ (mRNA level ⬇)	Protein expression ⬇ (mRNA level no change)

Note: CdR represents coding region; miRNA and mRNA are underlined to highlight the difference

1. An AMO is designed to entirely base-pair with the sequence of the target miRNA, whereas a miR-Mask is designed based on the sequence of the target site for a miRNA in the 3′UTR of the target mRNA. In other words, miR-Mask acts like a protector of the gene from being inhibited by miRNA.
2. An AMO interacts with (binds to) its target miRNA and may well cause degradation of that miRNA such that all functions of that miRNA are deemed to be eliminated, whereas a miR-Mask interacts with (binds to) its target mRNA and does not induce miRNA degradation such that the function of that miRNA on other genes is intact. In this sense, a miR-Mask is not only a target protector but also a miRNA protector.
3. The action of an AMO is miRNA-specific but not gene-specific and it may well induce enhancement of expression of multiple genes regulated by the same target miRNA, whereas a miR-Mask is expected to be gene-specific because it is fully complementary to the target mRNA sequence and is miRNA-specific as well because it is designed to target the binding site of that miRNA. Hence, a miR-Mask is expected to derepress only the target gene.
4. An AMO acts to disrupt miRNA:mRNA interaction by creating an AMO:miRNA interaction, whereas a miR-Mask disrupts miRNA:mRNA interaction by creating an ASO:mRNA interaction (ASO: anti-mRNA antisense oligomer).

The miR-Mask approach also defers from the conventional antisense technique in the following two aspects, despite that they are both entirely complementary to the target sequences.

1. A conventional antisense oligodeoxynucleotide (ASO) can in theory be designed to target any part of the protein-coding region of a gene (though the sequences from the translation start codon are frequently used), whereas a miR-Mask is limited to the target site of a miRNA in the 3′UTR of a protein-coding gene.
2. The efficacy of a miR-Mask on gene expression depends on the basal activity of the endogenous miRNA on the target mRNA, while that of a conventional ASO depends on the interaction between the ASO and the target gene.
3. A conventional antisense ODN binds to its target site in the coding region of a gene and hinders the protein translation process. Conversely, a miR-Mask binds to the 3′UTR and masks the target site of a miRNA to block the action of the endogenous miRNA and enhance protein translation. Thus, the two techniques produce exactly opposite outcomes: one inhibits but the other enhances gene expression.

A comparison of the miR-Mask, the AMO and the conventional antisense ODN techniques is summarized in Fig. 10.1.

10.4 Applications

The miR-Mask technology is an alternative to the AMO approach. But unlike AMO that acts in a non-gene-specific manner, the miR-Mask finds its particular value in

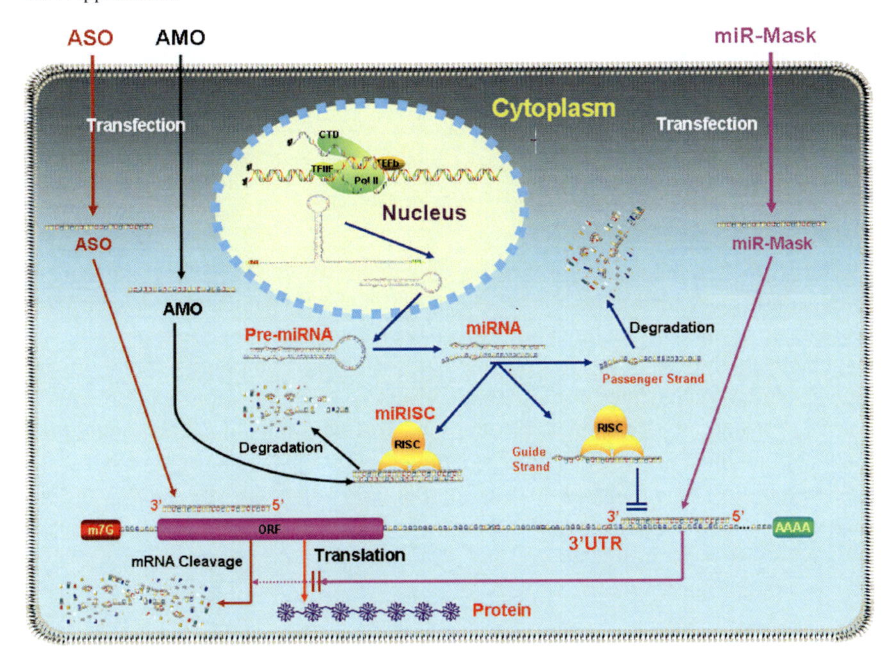

Fig. 10.1 Schematic presentation of actions of miRNA-masking antisense oligonucleiotide (miR-Mask) compared with the conventional antisense oligodeoxynucleotide (ASO) and anti-miRNA antisense inhibitor oligonucleotide (AMO) technologies. Synthetic nucleic acids are introduced into the cells. ASO binds to the coding region of the target mRNA and hinders the translation process; AMOs bind to the target miRNA, resulting in miRNA cleavage; miR-Masks bind to the binding site of miRNAs in 3′UTR of the target mRNA and prevent miRNAs from binding to the target mRNA, leading to a relief of translational repression without affecting miRNAs.

targeting miRNA in a gene-specific fashion. It is particularly useful when inhibiting miRNA action on a particular protein-coding gene without affecting the level of this miRNA and its silencing effects on other genes is retained.

We have validated the miR-Mask technology by testing its application to the cardiac pacemaker channel-encoding genes HCN2 and HCN4 (Xiao et al. 2007b). We created miR-Masks that are able to bind to HCN2 and HCN4 and prevent the repressive actions of miR-1 and miR-133. These miR-Masks resulted in enhanced protein expression of the pacemaker channels and increased pacemaker activities revealed by whole-cell patch-clamp recordings. Functionally, the miR-Masks for HCN channels cause acceleration of heart rate in rats, simulating "biological pacemakers" (Xiao et al. 2007b).

This technology has also been validated by a recent study in which the authors investigated the role of zebrafish miR-430 in regulating expression of TGF-β Nodal agonist squint and antagonist lefty, the key regulators of mesendoderm induction and left-right axis formation (Choi et al. 2007). They designed miR-Masks, which they called target protector morpholinos, complementary to miRNA binding sites in target mRNAs in order to disrupt the interaction of specific miRNA:mRNA pairs.

Protection of squint or lefty mRNAs from miR-430 resulted in enhanced or reduced Nodal signaling.

10.5 Advantages and Limitations

The miR-Mask approach is a valuable supplement to the AMO technique; while AMO is indispensable for studying the overall function of a miRNA, the miR-Mask might be more appropriate for studying the specific outcome of regulation of the target gene by the miRNA. (1) The major advantage of this technology is that it offers a gene-specific miRNA-interfering strategy, which in many situations is highly desirable. (2) This expression-enhancing action of miR-Mask is unique and could have many applications. (3) The characteristic dual specificities (miRNA-specificity and gene-specificity) of miR-Mask may be particularly useful for the miRNA:mRNA interactions consequent to polymorphisms in the protein-coding genes that create new binding sites for miRNAs.

References

Choi WY, Giraldez AJ, Schier AF (2007) Target protectors reveal dampening and balancing of Nodal agonist and antagonist by miR-430. Science 318:271–274.

Xiao J, Luo X, Lin H, Xu C, Gao H, Wang H, Yang B, Wang Z (2007a) MicroRNA miR-133 represses HERG K^+channel expression contributing to QT prolongation in diabetic hearts. J Biol Chem 282:12363–12367.

Xiao J, Yang B, Lin H, Lu Y, Luo X, Wang Z (2007b) Novel approaches for gene-specific interference via manipulating actions of microRNAs: Examination on the pacemaker channel genes *HCN2* and *HCN4*. J Cell Physiol 212:285–292.

Chapter 11
Sponge miR-Mask Technology

Abstract Sponge miR-Mask technology combines the principle of actions of miRNA Sponge and miR-Mask technologies. Like a miR-Mask but unlike a miRNA Sponge, a Sponge miR-Mask does not directly interact with a miRNA but is designed to bind to the binding site of a miRNA seed family in the 3′UTR of all target mRNAs; like a miRNA Sponge but unlike a miR-Mask, a Sponge miR-Mask binds by a partial complementary mechanism only with its seed site 8 nts base-pairing to its target genes. In this way, a Sponge miR-Mask is able to block access of all members of a miRNA seed family of interest to their binding sites or a particular miRNA to their multiple binding sites in a gene to target the actions of all members of that miRNA seed family, leading to derepression of the proteins from the miRNA seed family. This technology was established by my laboratory in 2008 (unpublished observations). The Sponge miR-Mask technology belongs to the "targeting-miRNA" and "miRNA-loss-of-function" strategy. The miRNA Sponge technology complies with the 'miRNA Seed Family' concept (see Sect. 2.1.4 for detail).

11.1 Introduction

miRNAs can be classified into families by their seed sequences (5′-end 2–8 nts); miRNAs with the same seed site belong to the same seed family. The members of the same seed family likely have the same set of target genes or same cellular functions. In other words, a single target gene is regulated by multiple miRNAs sharing the same seed site. For instance, 19 miRNAs: miR-17-5p (Cloonan et al. 2008; Matsubara et al. 2007; Volinia et al. 2006), miR-20a-b, miR-93, miR-106(a-b) (Garzon et al. 2006), miR-198 (Zhao et al. 2009), miR-372 (Voorhoeve et al. 2006), miR-520a-e,g-h (Bentwich et al. 2005) and miR-519b-e (Bentwich et al. 2005) all have a seed sequence CAAAGUGCU or AAAGUGCU. The members of this seed family can all repress STAT3 protein. Moreover, a single target gene (mRNA)

Z. Wang, *MicroRNA Interference Technologies*,
DOI: 10.1007/978-3-642-00489-6_11, © Springer-Verlag Berlin Heidelberg 2009

may bear multiple binding sites in its 3′ UTR for a miRNA or a miRNA seed family.

In many circumstances, thorough inhibition of target gene repression by a miRNA is highly desirable. To achieve this goal, inhibition of a whole seed family may be required. However, none of the miRNAi technologies described in previous chapters gives us the capability of antagonizing all members of a seed family. The miRNA Sponge technology was developed for this purpose, as detailed in Chap. 9 (Ebert et al. 2007; Hammond 2007). Unfortunately, in some cases, miRNA Sponges, while being able to effectively antagonize the target miRNA when fully antisense to this miRNA, are ineffective in inhibiting the other members of the same seed family that have variations in their sequences at 3′-end regions. Moreover, we have experimentally excluded the effectiveness of a truncated AMO that has only 8 nts exactly antisense to the seed sequence of a target miRNA (eg. miR-1). These results clearly indicate that full length with full complementarity of an AMO toward its target miRNA is required for an effective knockdown of the miRNA. An AMO with partial complementarity or partial sequence (e.g., only the seed site) against a miRNA will not lead to effective knockdown of the target miRNA. This has actually been demonstrated (Vermeulen et al. 2007).

One approach to tackle this problem is to use the MT-AMO technology developed in our laboratory (see Chap. 8) (Lu et al. 2009). With this technology, we can incorporate multiple AMO units with each of them directed against one member of a seed family of interest.

Alternatively, if one requires removing the function of a seed family but keeping the target miRNAs intact, one can focus on the protein-coding genes (mRNAs), instead of on miRNAs. This concept is essentially the same as that of our miR-Mask strategy (Xiao et al. 2007; Choi et al. 2007). One can design an oligodeoxynucleotide fragment with is 5′-end 8 nts exactly the same as or equivalent to (to be more accurate), the seed sequence of a seed family under test and with the rest of the region devoid of any seed sites for other miRNAs. This new technology, we have named Sponge miR-Mask, unites the concepts of miRNA Sponge and miR-Mask into one entity. The Sponge miR-Mask technology acts to mask or protect the binding sites of a selected seed family in the 3′UTR of a target mRNA.

11.2 Protocols

1. Select a miRNA seed family to begin your study. For example, the miR-17-5p family including miR-17-5p, miR-20(a-b), miR-93, miR-106(a-b), miR-198, miR-372, miR-520(a-e,g-h) and miR-519(b-e) (see Fig. 11.1);
2. Design an oligodeoxyribonucleotide (ODN) fragment exactly the same as the seed sequence 5′-AAAGUGCU-3′ of this seed family: 5′-**AAAGTGCT**-3′;
3. Then attach to the 3′-end of this fragment a 22-nt universal ODN fragment to form a 30-nt Sponge miR-Mask: Seed – ACTTTATCTATCTATTTATCGG.

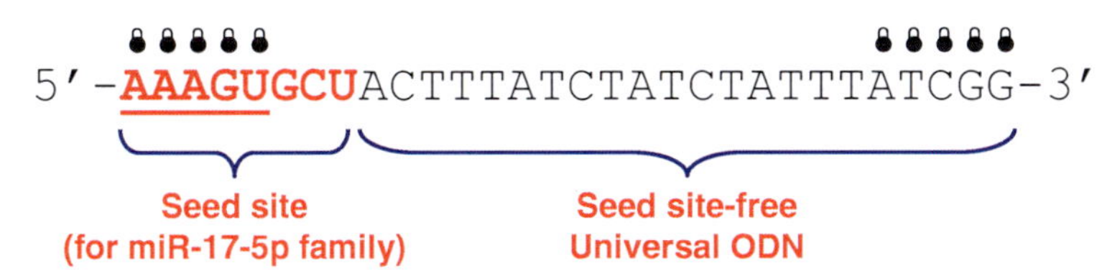

Fig. 11.1 Diagram illustrating the construction of a Sponge miR-Mask. To enhance the stability and affinity, the Sponge miR-Mask is chemically modified to have 5 nts at both ends, locked with methylene bridges (LNA)

For the miR-17-5p seed family, we have **AAAGTGCT** ACTTTATCTATCTATT-TATCGG. This universal ODN does not contain any seed sequences for other miRNAs;

4. Chemical synthesize the Sponge miR-Mask using services provided by commercial companies such as IDT Technologies or Ambion. Remember to chemical modify the oligonucleotide as described in Chap. 6. We found that LNA modification yields high affinity to the binding sites and high stability in cells (unpublished observations);
5. Transfect the Sponge miR-Mask into cells to study the enhancing effects on, or upregulation of, protein expression of the expected target genes (e.g., STAT3).

11.3 Principle of Actions

The Sponge miR-Mask technology integrates the strategies of miRNA Sponge (Ebert et al. 2007; Hammond 2007) and miR-Mask Xiao et al. 2007,; Choi et al. 2007) to generate unique oligodeoxribonucleotide (ODN) fragments for interfering with miRNA function (Xiao et al. 2007; Choi et al. 2007). Each Sponge miR-Mask ODN contains two parts: 5′-end seed sequence of 8 nts and a 3′-end random sequence of 22 nts; the seed sequence varies depending on the miRNA seed family under study whereas the random sequence can be a universal fragment that is sure of bearing no seed sites for any miRNAs. When introduced into cells, a Sponge miR-Mask will bind with its seed sequence to any genes containing the corresponding binding sites. When binding, the seed site of a Sponge miR-Mask will entirely base-pair with the target gene but the universal fragment will only partially base-pair with the target gene. In this way it blocks the accessibility of endogenous miRNAs carrying the same seed sequence, the target miRNAs you selected for your study, to abrogate the actions of those miRNAs resulting in relief of repression of targeted genes (Fig. 11.2). Hence, the Sponge miR-Mask technology belongs to the

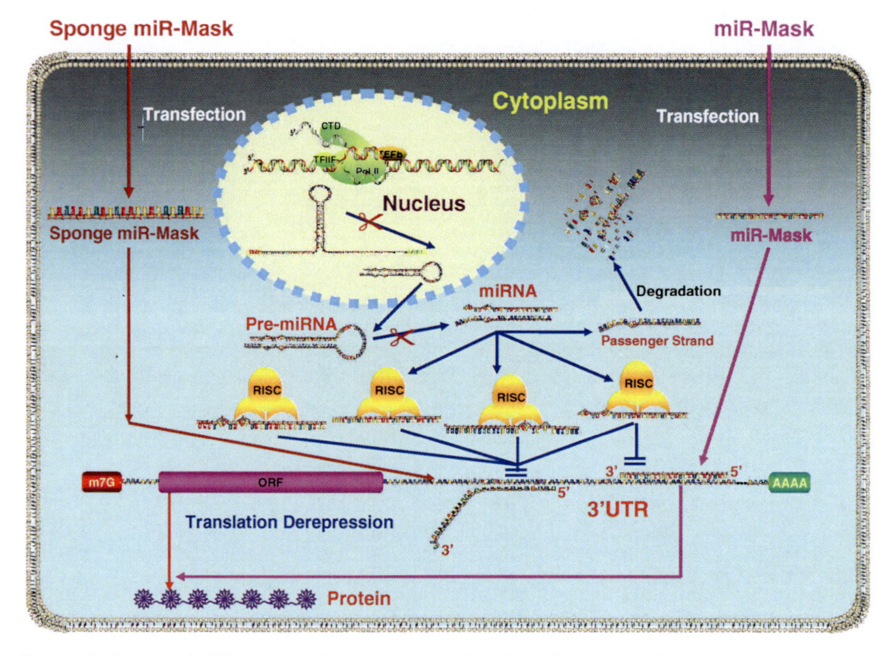

Fig. 11.2 Schematic illustration of the principle of action of Sponge miR-Mask, as compared to that of miR-Mask. A Sponge miR-Mask, once introduced into a cell (e.g., by transfection), binds with its seed sequence to the target site in the 3′UTR of the target gene. This binding blocks the access of all endogenous miRNAs belonging to a same seed family. By comparison, a miR-Mask blocks only a particular miRNA. The outcome is upregulation of expression of the target gene at the protein level due to derepression of the translation process

"targeting miRNA" and "miRNA-loss-of-function" strategy. On the other hand, it targets multiple members of a same seed family, thus it is miRNA seed-specific but not miRNA-specific. Additionally, as far as the target gene is concerned, the Sponge miR-Mask technology is not gene-specific.

Sponge miR-Mask and miRNA Sponge share three similarities (Table 11.1).

1. They both are designed to target the members of same miRNA seed families.
2. They both result in relief of repression or upregulation of the target protein-coding genes.
3. Moreover, they both are non-gene-specific but miRNA seed-specific.

Sponge miR-Mask differs from miRNA Sponge in three aspects.

1. A miRNA Sponge is antisense to its target miRNA and can bind to this miRNA (AMO:miRNA interaction), whereas a Sponge miR-Mask carries the same seed sequences of the target miRNAs or is antisense to the binding sites of the target miRNAs in the 3′UTR of the protein-coding genes (mRNA); they therefore bind to the protein-coding genes (ASO:mRNA interaction), not the target miRNAs.
2. A miRNA Sponge contains multiple identical AMO units, while a Sponge miR-Mask contains only one seed site.

Table 11.1 Comparison of three different miRNAi technologies

	miRNA Sponge	miR-Mask	Sponge miR-Mask
Target Interaction	miRNA	mRNA carrying the binding site of an miRNA	mRNA carrying the binding sites of a miRNA or a miRNA seed family
Complementarity	Seed site (8 nts) complementarity	Full (22 nts) complementarity	Seed site (8 nts) complementarity
Targeted miRNA	miRNA degradation	miRNA intact	miRNA intact
Specificity	miRNA seed family-specific	miRNA-specific mRNA-specific	miRNA seed family-specific
Outcome	Derepression of proteins	Derepression of proteins	Derepression of proteins

3. A miRNA Sponge may cause degradation of targeted miRNAs, whereas a Sponge miR-Mask keeps both targeted miRNAs and protein-coding genes intact.

Comparing with a miR-Mask, a Sponge miR-Mask also has similarities and differences. The similarities include the following points (Table 11.1).

1. Both a miR-Mask and a Sponge miR-Mask directly interact with target protein-coding genes by binding to sites of endogenous miRNAs in the 3′UTRs of the genes.
2. Both a miR-Mask and a Sponge miR-Mask upregulate expression of protein-coding genes by interrupting the miRNA:mRNA interaction.
3. Both a miR-Mask and a Sponge miR-Mask are anti-mRNA antisense ODNs.

The differences between a miR-Mask and a Sponge miR-Mask are indicated by the following aspects.

1. A miR-Mask is designed to bind to a particular site of a selected target gene by full base-pairing; it is therefore gene-specific. By comparison, a Sponge miR-Mask is designed to interact with multiple binding sites for a miRNA or a seed family by a partial complementary mechanism. The Sponge miR-Mask technology is thus non-gene-specific and is able to target any genes carrying the binding sites for that miRNA seed family.
2. A miR-Mask may carry more than one seed sequence. For instance, in the 3′UTR of human KCND3 gene (encoding Kv4.3 voltage-dependent K^+channel subunit), there is a region containing the sequence: 5′-AAACCACTGGACAGAGGGC-CAG-3′, where it carries binding sites for the seed sequences of three different miRNAs. "AGGGCCAG" contains the binding sequence for miR-328, "CACTGGAC" for miR-145 and "CAGUGGUU" for miR-140. If one designs a miR-Mask for miR-328, then the fragment actually covers miR-140 and miR-145 and can possibly affect the effects of these two non-target miRNAs. To avoid this undesirable effect, one must search for other parts of the 3′UTR sequence to see if the miR-328 binding site is available for creating a miR-Mask. However,

if one is to use a Sponge miR-Mask, then one will not have this problem because a Sponge miR-Mask is able to mask only one binding site.

3. A miR-Mask can normally act to protect only one binding site of a miRNA, whereas a Sponge miR-Mask is able to act on all binding sites for a miRNA seed family existing in the 3′UTR of a protein-coding gene. The effect of a Sponge miR-Mask is expected to be more thorough in case a gene carries multiple binding sites for a miRNA.

11.4 Applications

The Sponge miR-Mask technology can be used to target a whole miRNA seed family aiming at upregulating gene expression of genes at the protein level. We used this approach to establish STAT3 as a target gene for and the consequent proapoptotic actions of, the miR-17-5p seed family in cardiac cells (unpublished observations). In response to oxidative stress, the members of the miR-17-5p seed family are upregulated in their expression. This upregulation results in repression of STAT3 and apoptotic cell death. Application of a Sponge miR-Mask to the cells abrogated the repression of STAT3 and apoptosis by 85 and 77%, respectively. By comparison, application of a miRNA Sponge toward miR-17-5p produced much smaller effects: 33 and 24% reduction of repression of STAT3 and apoptosis, respectively. This is explained by the fact that the miRNA Sponge is effective in antagonizing only miR-17-5p and much less effective in inhibiting other members of the seed family. Blockade of miR-17-5p does not prevent the binding of other members of the seed family to STAT3 to elicit the repressive effect. Similarly, a miR-Mask was also found to be less effective in reversing the repression of STAT3 and apoptosis (41 and 32%, respectively), compared with the Sponge miR-Mask. This is because there are three binding sites in the 3′UTR of STAT3 and protection of only one site by the miR-Mask is insufficient to abolish the post-transcriptional repression of STAT3 by endogenous miR-17-5p seed family members. By comparison, the Sponge miR-Mask is able to protect all three binding sites to produce greater masking effects. In such a case, the Sponge miR-Mask demonstrates its superiority over other miRNAi technologies.

11.5 Advantages and Limitations

The major advantage of the Sponge miR-Mask approach is its ability to simultaneously relieve the repressive actions of all members of a same seed family, an effect highly desirable under many situations. This goal could hardly be achieved by any other miRMAi technologies.

One obvious limitation of the Sponge miR-Mask technology is associated with its poor gene-specificity; it is specific to all genes carrying the same seed binding site but not to a particular gene. Taking our Sponge miR-Mask for the miR-17-5p seed family as an example, the Sponge miR-Mask is expected to not only act on STAT3 but also on any other gene carrying the binding sites for the miR-17-5p seed family. This can also produce unwanted actions.

References

Bentwich I, Avniel A, Karov Y, Aharonov R, Gilad S, Barad O, Barzilai A, Einat P, Einav U, Meiri E, Sharon E, Spector Y, Bentwich Z (2005) Identification of hundreds of conserved and nonconserved human microRNAs. Nat Genet 37:766–770.

Choi WY, Giraldez AJ, Schier AF (2007) Target protectors reveal dampening and balancing of Nodal agonist and antagonist by miR-430. Science 318:271–274.

Cloonan N, Brown MK, Steptoe AL, Wani S, Chan WL, Forrest AR, Kolle G, Gabrielli B, Grimmond SM (2008) The miR-17-5p microRNA is a key regulator of the G1/S phase cell cycle transition. Genome Biol 9:R127.

Ebert MS, Neilson JR, Sharp PA (2007) MicroRNA sponges: Competitive inhibitors of small RNAs in mammalian cells. Nat Methods 4:721–726.

Garzon R, Pichiorri F, Palumbo T, Iuliano R, Cimmino A, Aqeilan R, Volinia S, Bhatt D, Alder H, Marcucci G, Calin GA, Liu CG, Bloomfield CD, Andreeff M, Croce CM (2006) MicroRNA fingerprints during human megakaryocytopoiesis. Proc Natl Acad Sci USA 103:5078–5083.

Hammond SM (2007) Soaking up small RNAs. Nat Methods 4:694–695.

Lu Y, Xiao J, Lin H, Bai Y, Luo X, Wang Z, Yang B (in press) A single anti-microRNA antisense oligodeoxynucleotide (AMO) targeting multiple microRNAs offers an improved approach for microRNA interference. Nucleic Acids Res.

Matsubara H, Takeuchi T, Nishikawa E, Yanagisawa K, Hayashita Y, Ebi H, Yamada H, Suzuki M, Nagino M, Nimura Y, Osada H, Takahashi T (2007) Apoptosis induction by antisense oligonucleotides against miR-17-5p and miR-20a in lung cancers overexpressing miR-17-92. Oncogene 26:6099–6105.

Vermeulen A, Robertson B, Dalby AB, Marshall WS, Karpilow J, Leake D, Khvorova A, Baskerville S (2007) Double-stranded regions are essential design components of potent inhibitors of RISC function. RNA 13:723–730.

Volinia S, Calin GA, Liu CG, Ambs S, Cimmino A, Petrocca F, Visone R, Iorio M, Roldo C, Ferracin M, Prueitt RL, Yanaihara N, Lanza G, Scarpa A, Vecchione A, Negrini M, Harris CC, Croce CM (2006) A microRNA expression signature of human solid tumors defines cancer gene targets. Proc Natl Acad Sci USA 103:2257–2261.

Voorhoeve PM, le Sage C, Schrier M, Gillis AJ, Stoop H, Nagel R, Liu YP, van Duijse J, Drost J, Griekspoor A, Zlotorynski E, Yabuta N, De Vita G, Nojima H, Looijenga LH, Agami R (2006) A genetic screen implicates miRNA-372 and miRNA-373 as oncogenes in testicular germ cell tumors. Cell 124:1169–1181.

Xiao J, Yang B, Lin H, Lu Y, Luo X, Wang Z (2007) Novel approaches for gene-specific interference via manipulating actions of microRNAs: Examination on the pacemaker channel genes *HCN2* and *HCN4*. J Cell Physiol 212:285–292.

Zhao JJ, Yang J, Lin J, Yao N, Zhu Y, Zheng J, Xu J, Cheng JQ, Lin JY, Ma X (2009) Identification of miRNAs associated with tumorigenesis of retinoblastoma by miRNA microarray analysis. Childs Nerv Syst 25:13–20.

Chapter 12
miRNA Knockout Technology

Abstract The miRNA Knockout (miR-KO) technology aims to generate mouse lines with genetic ablation of specific miRNAs or targeted disruption of miRNA genes. This approach allows for investigations of miRNA function related to the development of particular biological processes and/or pathological conditions in an in vivo context and in a permanent setting. For some applications, this knockout strategy has demonstrated its superiority over knockdown strategies such as antisense to miRNA, which are primarily in vitro, transient, local targeting-miRNA or miRNA-loss-of-function approaches. The first applications of knockout models to study miRNA function in development were performed in flies in 2005 by Kwon C, Han Z, Olson EN, Srivastava D [Proc Natl Acad Sci USA 102:18986–18991, 2005] and by Sokol and Ambros [Genes Dev 19:2343–2354, 2005]. Later, the miR-KO techniques were introduced to mouse models by Zhao et al. [Cell 129:303–311, 2007] and Thai et al. [Science 316:604–608, 2007]. It seems that this strategy is increasingly appreciated and favored by recent studies [Kuhnert F, Mancuso MR, Hampton J, Stankunas K, Asano T, Chen CZ, Kuo CJ, Development 135:3989–3993, 2008; Wang S, Aurora AB, Johnson BA, Qi X, McAnally J, Hill JA, Richardson JA, Bassel-Duby R, Olson EN, Dev Cell 15:261–271, 2008], though important limitations exist.

12.1 Introduction

The Targeting-miRNA or miRNA-loss-of-function technologies described in the preceding chapters are primarily in vitro, transient, local miRNA knockdown strategies, despite that the antagomiR approach offers an opportunity for reasonably long-lasting, global loss-of-function of specific miRNAs. Often, to understand thoroughly the function of miRNAs in the development of biological processes and pathological conditions, it is highly desirable to have an efficient means of disrupting miRNA genes in an in vivo context and in a permanent and tissue-specific

Z. Wang, *MicroRNA Interference Technologies*,
DOI: 10.1007/978-3-642-00489-6_12,

manner. Targeted miRNA deletion or miRNA-gene knockout (miR-KO) techniques meet the requirements for this type of study.

There are in general two ways to disrupt the expression of a protein-coding gene: targeted homologous recombination or the insertion of gene trap cassettes. Similar approaches can be applied to miRNA genes.

Intriguingly, in light of the fact that currently about half of the genes in a mouse have been knocked out and over 50% of known miRNAs are located within introns of coding genes, a study investigated the possibility that intronic miRNAs may have been coincidentally deleted or disrupted in some of these mouse models (Osokine et al. 2008). The authors searched published murine knockout studies and gene trap embryonic stem cell line databases for cases where a miRNA was located within or near the manipulated genomic loci. They found almost 200 cases where miRNA expression may have been disrupted along with another gene. The results draw attention to the need for careful planning in future knockout studies to minimize the unintentional disruption of miRNAs. These data also raise the possibility that many knockout studies may need to be re-examined to determine if loss of a miRNA contributes to the phenotypic consequences attributed to loss of a protein-encoding gene.

12.2 Protocols

12.2.1 Homologous Recombination Methods

The neomycin (Neo) resistance cassette using homologous recombination method has been mostly used for generating specific miRNA knockout mice. It has been used for miR-1 (Zhao et al. 2007), miR-126 (Wang et al. 2008; Kuhnert et al. 2008), miR-223 (Johnnidis et al. 2008) and miR-155 (Thai et al. 2007).

1. Select a miRNA to be targeted;
2. PCR amplify a fragment spanning the target miRNA from the mouse genomic DNA or use restriction enzymes to cut out a fragment covering the target miRNA.

For example, in a study reported by Johnnidis et al. (2008), 5′ and 3′ sequences flanking the endogenous 110-bp miR-223 locus on the X chromosome were amplified by PCR from a C57BL/6 genomic BAC clone (BACPAC Resource Center), generating 6.8-kb and 1.5-kb fragments, respectively. In the study reported by Wang et al. (2008), a 5.7kb fragment (5′ arm) extending upstream of the miR-126 coding region and a 1.8-kb fragment (3′ arm) downstream of the miR-126 coding region was obtained;

3. Clone these homology arms into a vector incorporating both a neomycin resistance cassette for positive selection and a diphtheria toxin (DTA) gene for negative selection. Alternatively, they can be cloned into the pGKneoF2L2dta

targeting plasmid upstream and downstream of the loxP sites and the Frt-flanked neomycin cassette, respectively;
4. Linearize the targeting vector and transfect it into V6.5 embryonic stem (ES) cells by electroporation (Eggan et al. 2001) or by liposome reagent;
5. Isolate recombinant ES clones after culturing in medium containing G418 antibiotic and screen for proper integration by means of PCR amplification of the interval between the neo cassette and the junction of the short 3′ homology arm with downstream genomic sequence. Sequencing: verify the sequence;
6. Micro-inject the positive colonies into 3.5 day post-coitus blastocysts to generate high-percentage chimeras that are to be bred to recover heterozygous mice with germline transmission;
7. Cross the resulting chimeric offspring (*miRNAneo/+*mice) to C57BL/6 mice (or to CAG-, CMV- or HPRT-Cre transgenic mice) to obtain the mutant miRNA allele;
8. Backcross the male offspring for four generations to C57BL/6 mice congenic for CD45.1 (Figs. 12.1 and 12.2).

12.2.2 Cre-loxP Methods

A brief description of loxP methods is given below based on the study reported by Kuhnert et al. (2008).

1. Select a target miRNA for study and obtain the gene encoding this miRNA;
2. Analyze the sequence flanking the precursor miRNA to identify the restriction sites for NheI at 5′- and NsiI at 3′-regions;
3. Clone a loxP site (Pl452) and a neomycin selection cassette plus a loxP site (Pl451) into an NheI site 5′ of the miRNA and an NsiI site 3′ of the miRNA, respectively;
4. Generate delta alleles by crossing to CMV- or HPRT-Cre mice;
5. Analyze the mutant mice in a mixed 129sV/C57Bl/6 genetic background.

12.2.3 FLP-FRT Deletion Methods

Below is an introduction of *Drosophila* miRNA-1 (dmiR-1) locus deletion using piggyBac insertion lines, reported in the studies of Kwon et al. (2005) and Sokol and Ambros (2005).

1. Select transposon elements flanking the region (target miRNA) to be deleted;
2. Maintain fly stocks at 22 or 25°C on standard media;
3. Let males carrying one element with females carrying a FLP recombinase transgene;

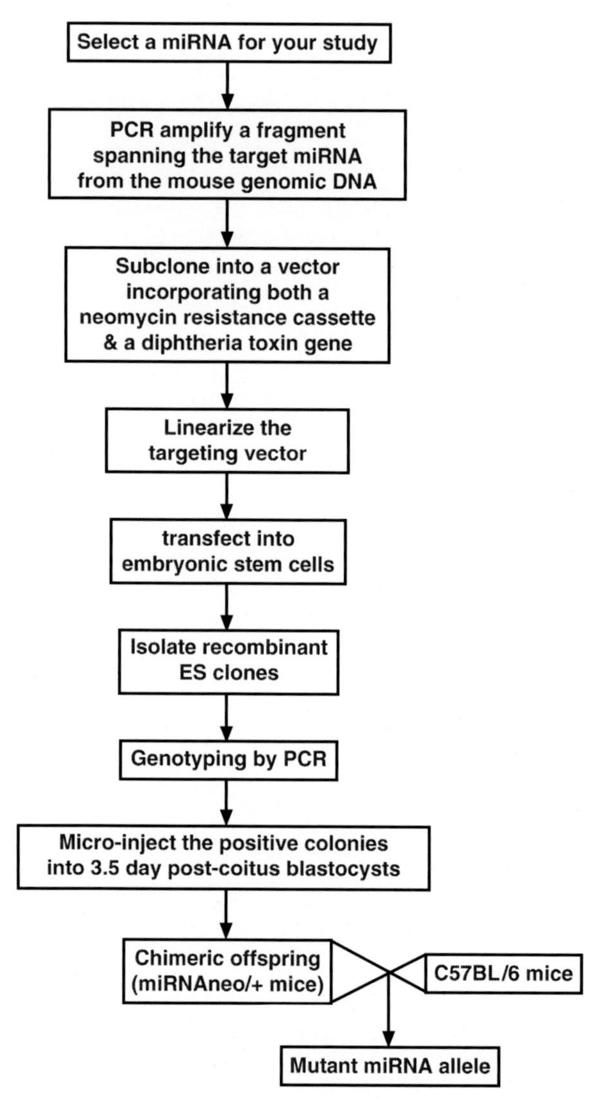

Fig. 12.1 Flowchart of using homologous recombination methods to generate miRNA knockout (miR-OK) mice

4. Then let progeny males carrying both the element and FLP recombinase mate to females carrying the second element;
5. After two days, subject parents and progeny (progeny contain both the FRT-bearing elements in *trans* and FLP recombinase) to a 1 h heat shock by placing the bottles into a 37°C water bath. Then remove parents after 72 h of total egg-laying time and subject the bottles to daily 1 h heat shocks for 4 more days;

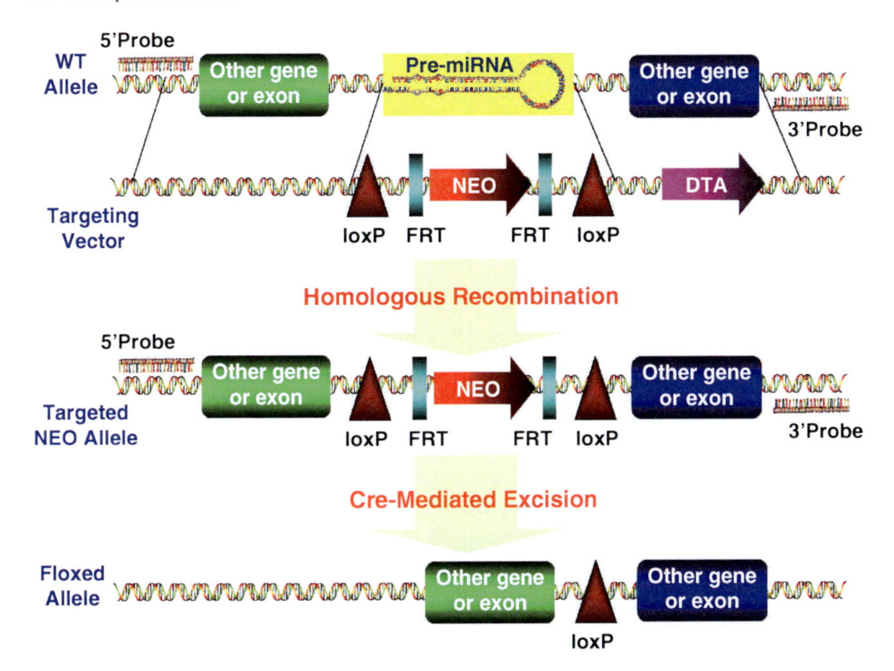

Fig. 12.2 Strategy to generate miRNA mutant mice by homologous recombination, based on the study reported by Wang et al. (2008). The pre-miRNA sequence in the mouse genomic region is replaced with a neomycin resistance cassette (NEO) flanked by loxP sites. NEO is removed in the mouse germline by crossing heterozygous mice to CAG-Cre transgenic mice. DTA:diphtheria toxin A

6. Raise progeny to adulthood, collect virgin females and cross them to males containing marked balancer chromosomes;
7. Cross individual progeny males (five *w* – males for a *w* – deletion; 50 males for a *w* + deletion) pairwise to virgin females to generate additional progeny for PCR confirmation analysis and to balance the stocks in an isogenic background;
8. PCR confirm FLP-FRT-based deletions using purified genomic DNA from homogenized flies from each isolate line. In general, it is sufficient to test 5 putative lines for each *w* – deletion and 50 putative lines for each *w* + deletion.

12.3 Principle of Actions

The miR-KO silences expression of a particular miRNA with concomitant derepression of the target protein-coding genes of that miRNA, via a complete, miRNA-specific and may be temporal- and tissue-specific, genetic ablation of the target miRNA.

12.4 Applications

miR-KO technology has increasingly found its value in miRNA research. More and more studies using this approach are appearing in the literature. Its future application to the field is expected to be more and more appreciated.

1. To acquire a conclusive evidence for the role of a target miRNA through permanent removal of a target miRNA.
2. To allow for controllable or conditional miRNA silencing (Vong et al. 2005).
3. To allow for studying miRNA-loss-of-function at both in situ cellular level and in vivo whole-animal context.

miR-KO technology has been applied to studying the consequences of loss-of-function of particular miRNAs on a few occasions.

In *Drosophila*, deletion of the single miR-1 gene (dmiR-1), expressed specifically in cardiac and somatic muscle, results in a defect in muscle differentiation or maintenance (Kwon et al. 2005; Sokol and Ambros 2005).

To define the in vivo function of a specific miRNA in mammals, Zhau et al. (2007) targeted the miR-1-2 sequence for deletion by homologous recombination in mouse embryonic stem (ES) cells.

Targeted deletion of miR-126 in mice suggests that miR-126 functions as an endothelial cell-specific regulator of angiogenic signaling (Wang et al. 2008). The knockout model results in vascular leakage, hemorrhaging and embryonic lethality in a subset of mutant mice. These vascular abnormalities can be attributed to a loss of vascular integrity and defects in endothelial cell proliferation, migration and angiogenesis. The subset of mutant animals that survives is prone to cardiac rupture and lethality following myocardial infarction with defective vascularization of the infarct. The proangiogenic actions of miR-126 correlate with its repression of Spred-1, a negative regulator of MAP kinase signaling.

Another study explored the functions of both Egfl7 and its embedded miRNA, miR-126, using floxed alleles to selectively disrupt each gene without reciprocal perturbation (Kuhnert et al. 2008). The endothelial expressed Egfl7/miR-126 locus contains miR-126 within Egfl7 intron 7 and angiogenesis deficits have been previously ascribed to Egfl7 gene-trap and lacZ knock-in mice. Selectively floxed Egfl7 and miR-126 alleles revealed that Egfl7-floxed mice are phenotypically normal, whereas miR-126-floxed mice bearing a 289-nt microdeletion recapitulate previously described Egfl7 embryonic and postnatal retinal vascular phenotypes. Regulation of angiogenesis by miR-126 was confirmed by endothelial-specific deletion and in the adult cornea micropocket assay. Furthermore, miR-126 deletion inhibits VEGF-dependent Akt and Erk signaling by derepression of the p85β subunit of PI3 kinase and of Spred1, respectively. These studies demonstrate the regulation of angiogenesis, by an endothelial miRNA, attribute previously described Egfl7 vascular phenotypes to miR-126 and document the inadvertent disruption of miRNA expression by conventional deletion and gene-trap knockout approaches in mice, as recently predicted in a bioinformatics analysis by McManus and colleagues (Osokine et al. 2008).

Using genetic deletion in conjunction with transgenic approach, Thai et al. (2007) showed that the evolutionarily conserved miR-155 has an important role in the mammalian immune system, specifically in regulating T helper cell differentiation and the germinal center reaction to produce an optimal T cell–dependent antibody response. miR-155 exerts this control, at least in part, by regulating cytokine production.

Johnnidis et al. (2008) reported that the myeloid-specific miR-223 negatively regulates progenitor proliferation and granulocyte differentiation and activation, acting as a fine-tuner of granulocyte production and the inflammatory response. miR-223 mutant mice have an expanded granulocytic compartment resulting from a cell autonomous increase in the number of granulocyte progenitors. They further identified MEF2c, a transcription factor that promotes myeloid progenitor proliferation, is a target of miR-223 and that genetic ablation of MEF2c suppresses progenitor expansion and corrects the neutrophilic phenotype in miR-223 null mice. In addition, granulocytes lacking miR-223 are hypermature, hypersensitive to activating stimuli and display increased fungicidal activity. As a consequence of this neutrophil hyperactivity, miR-223 mutant mice spontaneously develop inflammatory lung pathology and exhibit exaggerated tissue destruction after endotoxin challenge.

Nonetheless, it should be noted that miR-KO will hardly be used for therapeutic purpose.

12.5 Advantages and Limitations

Genetically modified mouse models in which a specific gene is removed or replaced have proven to be powerful tools for identification/validation of target gene and scientific understanding of molecular mechanisms underlying gene expression regulation through mechanistic studies. If properly designed and appropriately applied, miRNA-knockout approach offers a most clean and complete loss of function of miRNAs. Yet in spite of all these advantages, there are significant limitations of genetically modified mouse models.

1. One drawback to disrupt a single miRNA gene is that many miRNAs are members of a family of closely related miRNAs encoded by more than one gene located on different chromosomes and often display redundancy in the primary coding sequence with the functions of which being redundant but yet can be differentially regulated at the transcriptional and post-transcriptional levels. This makes genetic knockout of miRNAs difficult, if not impossible.
2. Another challenge in phenotypic discovery using miRNA gene disruption techniques is the unknown manner in which multiple miRNAs interact to carry out functions in the cell.
3. Modification of a given gene does not always result in the anticipated phenotype. In some instances, phenotypes of targeted mouse mutants may not be those predicted from the presumed function of the given genes, while other null mutants demonstrate no apparent defects.

4. Furthermore, the phenotypic outcome can be influenced by many environmental and genetic factors. Therefore, interpretation of the significance of the findings from studies using genetically modified mouse models is not always as straightforward as one would expect, especially when desire is to extrapolate the findings to humans.
5. Compared to available miRNA-knockdown approaches, miRNA-knockout is a relatively difficult, complicated, time-consuming, costly procedure.

References

Eggan K, Akutsu H, Loring J, Jackson-Grusby L, Klemm M, Rideout WM 3rd, Yanagimachi R, Jaenisch R (2001) Hybrid vigor, fetal overgrowth, and viability of mice derived by nuclear cloning and tetraploid embryo complementation. Proc Natl Acad Sci USA 98:6209–6214.

Johnnidis JB, Harris MH, Wheeler RT, Stehling-Sun S, Lam MH, Kirak O, Brummelkamp TR, Fleming MD, Camargo FD (2008) Regulation of progenitor cell proliferation and granulocyte function by microRNA-223. Nature 451:1125–1129.

Kuhnert F, Mancuso MR, Hampton J, Stankunas K, Asano T, Chen CZ, Kuo CJ (2008) Attribution of vascular phenotypes of the murine Egfl7 locus to the microRNA miR-126. Development 135:3989–3993.

Kwon C, Han Z, Olson EN, Srivastava D (2005) MicroRNA1 influences cardiac differentiation in Drosophila and regulates Notch signaling. Proc Natl Acad Sci USA 102:18986–18991.

Osokine I, Hsu R, Loeb GB, McManus MT (2008) Unintentional miRNA ablation is a risk factor in gene knockout studies: A short report. PLoS Genet 4:e34.

Sokol NS, Ambros V (2005) Mesodermally expressed Drosophila microRNA-1 is regulated by Twist and is required in muscles during larval growth. Genes Dev 19:2343–2354.

Thai TH, Calado DP, Casola S, Ansel KM, Xiao C, Xue Y, Murphy A, Frendewey D, Valenzuela D, Kutok JL, Schmidt-Supprian M, Rajewsky N, Yancopoulos G, Rao A, Rajewsky K (2007) Regulation of the germinal center response by microRNA-155. Science 316:604–608.

Vong LH, Ragusa MJ, Schwarz JJ (2005) Generation of conditional Mef2cloxP/loxP mice for temporal- and tissue-specific analyses. Genesis 43:43–48.

Wang S, Aurora AB, Johnson BA, Qi X, McAnally J, Hill JA, Richardson JA, Bassel-Duby R, Olson EN (2008) The endothelial-specific microRNA miR-126 governs vascular integrity and angiogenesis. Dev Cell 15:261–271.

Zhao Y, Ransom JF, Li A, Vedantham V, von Drehle M, Muth AN, Tsuchihashi T, McManus MT, Schwartz RJ, Srivastava D (2007) Dysregulation of cardiogenesis, cardiac conduction, and cell cycle in mice lacking miRNA-1-2. Cell 129:303–311.

Chapter 13
Dicer Inactivation Technology

Abstract Targeting miRNAs via interrupting the miRNA biogenesis pathway has become an important alternative to directly targeting mature miRNAs. In theory, every single one of the components along the miRNA biogenesis could be manipulated to disrupt the pathway. Dicer, which is critical for processing of pre-miRNAs into their mature form, has been a popular target helping to assess the global requirement for miRNAs in mammalian biology. Numerous studies using the Dicer inactivation approach have been documented. While this approach provides an invaluable means of studying the global requirement of miRNA for certain biological and pathophysiological processes, it has its inherent limitations. The Dicer inactivation technology is in general a miRNA-loss-of-function strategy; nonetheless, as will be pointed out, it can give rise to paradoxical gain-of-function outcomes.

13.1 Introduction

An alternative to the miR-KO or targeted miRNA knockdown described in previous chapters is to disrupt the biogenesis pathway of miRNAs.

DGCR8 is an RNA-binding protein that assists the RNase III enzyme Drosha in the processing of miRNAs, a subclass of small RNAs. A Dgcr8 knockout model has been created to study the role of miRNAs in ES cell differentiation by generating. Analysis of mouse knockout ES cells shows that DGCR8 is essential for biogenesis of miRNAs (Wang et al. 2007).

Deficiency in Ago2 impairs miRNA biogenesis from precursor miRNAs followed by a reduction in miRNA expression levels. In mice, disruption of Ago2 leads to embryonic lethality early in development after the implantation stage (Morita et al. 2007). Another study showed that Ago2 controls early development of lymphoid and erythroid cells (O'Carroll et al. 2007).

Specific deletion of Drosha throughout the T cell compartment results in spontaneous inflammatory disease and autoimmunity (Chong et al. 2008).

Z. Wang, *MicroRNA Interference Technologies*,
DOI: 10.1007/978-3-642-00489-6_13,

Even though, a majority of studies taking the approach of miRNA biogenesis disruption targets Dicer. Dicer, being critical for processing of pre-miRNAs into their mature form, has become the most popular and efficient target for studying the global requirement of miRNA for various biological processes. The expression of Dicer in several adult tissues demonstrated that Dicer is strongly expressed in the heart, liver and kidney (Yang et al. 2005), suggesting that Dicer may have a more specialized function in these organs. Lower levels of Dicer expression are observed in the brain, spleen, lung, skeletal muscle and testis.

13.2 Protocols

Dicer is an RNase III; all currently available gene knockdown and knockout techniques could be applied to render Dicer inactivation. These include knockdown techniques such as siRNA and antisense and knockout techniques such as homologous recombination and gene insertion. The procedures for establishing Dicer knockout mouse models are essentially the same as those already described in Chap. 12 for miRNA knockout. Below, I give an example of Dicer knockout using the homologous recombination methods.

13.2.1 Neomycin-Expression Cassette Methods (Homologous Recombination)

1. Digest the mouse CITB BAC library BamHI-XhoI to obtain a fragment of ~10-kb long. Then ligate this fragment into a pBluescript plasmid (Stratagene);
2. Replace the 1.2-kb HindIII-PmlI fragment containing exons 1 and 2 of Dicer by a neo expression cassette to construct a Dicer-targeting vector;
3. Introduce the vector into Bruce-4 ES cells derived from C57BL/6 mice to establish Dicer$^{neo/+}$ ES cell clone as described (Kanellopoulou et al. 2005). Verify recombination by PCR assays of DNA from tail biopsies, using a 3′-flanking probe, a 5′-flanking probe and a neo probe. Both the 3′- and 5′-probes can be generated by PCR reaction off the BAC clone. The PCR primers are 3′-probe: 5′-TCTTCGTCGAAACGTACAAG-3′ and 5′-TGGAAAGACCCTCATTCC-AAG-3′ and 5′-probe: 5′-GTTATTACCACTAAATATCACG-3′ and 5′-CT-GCC-AAGGCTTTGTTTCAC-3′. The neo probe is a 259-bp PstI fragment from a plasmid containing the neo cassette;
4. Then inject a heterozygous Dicer$^{neo/+}$ ES cell clone into blastocysts to derive chimeric mice and implant into pseudopregnant females;
5. The neomycin-resistance cassette (neo) is flanked by Flp recognition target sites and can be removed from the conditional Dicer allele by Flp recombinase. Thus, to generate Dicer$^{Flp/+}$offspring, a founder Dicer$^{neo/+}$chimeric mouse needs to be bred to a Flp deleter transgenic mouse (Rodriguez et al. 2000) to yield

heterozygous F1 offspring, which will be then intercrossed to produce +/+, +/−, −/− offspring;

6. Confirm genotypes of mice by PCR reaction using the primers KOF (5′-AGCCA-TCTCCCCAGAAGTCC-3′), which is the forward primer common for both the wild-type and the targeted allele, KOR2 (5′-CCAAAGAACGGAGCCGGTTG-3′) and KOR1 (5′-CGTGTAGGGTTCAGTCATTCGT-3′), which are designed for the amplification of the wild-type and the targeted alleles, respectively;
7. To generate T cell–specific KO mice, CD4cre transgenic mice (Srinivas et al. 2001) need to be bred to $Dicer^{Flp/+}$mice and progeny need to be intercrossed. Then, breed R26R-YFP mice to $Dicer^{Flp/+}$

13.2.2 Cre-loxP Methods

The detailed information regarding the knockout mice carrying the Dicer floxed allele has been described previously (Yi et al. 2006).

1. Flox exon 23 of the Dicer locus by two loxP loci (referred to as $Dicer^{Flox}$);
2. Let the $Dicer^{Flox/Flox}$ mice mate with Zp3-Cre transgenic mice, which express Cre recombinase under the control of the Zona pellucida glycoprotein 3 promoter (de Vries et al. 2000; Sohal et al. 2001);
3. Then let the $Dicer^{Flox/+}$/Zp3-Cre female mice mate with $Dicer^{Flox/Flox}$ male mice. This mating will generate $Dicer^{-/Flox}$/Zp3-Cre mice;
4. Following the deletion of the floxed allele in the oocyte, we generate oocytes that are the null mutants for Dicer.

13.3 Principle of Actions

The application of Dicer inactivation technology is fairly simple and straightforward: to disrupt miRNA biogenesis for exploring the role of Dicer or miRNA in certain biological or pathophysiological processes.

Disrupt the Dicer gene via homologous recombination in embryonic stem cells. Because exon 1 contains the putative translation initiation codon AUG, an elimination of the first two exons of the mouse dicer gene with a neo expression cassette is deemed to abolish the expression of a functional Dicer protein.

13.4 Applications

Dicer inactivation is a powerful approach to defining the global requirement of miRNAs in virtually any kind of biological process of organisms. It has the following applications in miRNA research.

1. To achieve a global loss-of-function of literally all cellular miRNAs.
2. To create permanent loss-of-function of the miRNA population.
3. To allow for controllable or conditional upregulation of protein-coding genes in a non-specific fashion.
4. To allow for studying miRNA-loss-of-function at both *in situ* cellular level and *in vivo* whole-animal context.

Application of the Dicer Inactivation Technology has been realized on a number of occasions. Below are a few examples to give readers an idea of what we know about miRNA function using this approach.

miRNAs are required for cell-lineage decisions (Gu et al. 2008). Dicer germline null allele results in embryonic lethality. Dicer inactivation homozygous embryos displayed a retarded phenotype and died between days 12.5 and 14.5 of gestation (Yang et al. 2005). To avoid this problem, a floxed Dicer allele was deleted using a myocardium-restricted, temporally regulated Cre deleter strain in one study (da Costa Martins et al. 2008). Inactivation of Dicer in the postnatal heart results in severe cardiac malfunction and failure regardless of the timing of gene deletion. The malfunction includes spontaneous cardiac remodeling, impairment of cardiac function and premature death within 1 week. Dicer ablation in 8-week-old mice provokes a remarkably rapid and spontaneous myocardial growth, accompanied by a severe histopathology. In the adult myocardium, loss of Dicer induced rapid and dramatic biventricular enlargement, accompanied by hypertrophic growth of cardiomyocytes, myofiber disarray, ventricular fibrosis, strong induction of fetal gene transcripts and functional defects. The apparent phenotypic difference between postnatal Dicer ablation in juvenile and adult myocardium likely reflects the different maturing states of the postnatal myocardium and could point toward relative differences between overall miRNA biogenesis and biological contribution in the juvenile versus adult myocardium. Clearly, Dicer depletion in the juvenile heart provoked an overall tendency toward arrhythmogenesis and less marked myocyte hypertrophy, whereas in the adult myocardium pronounced myocyte hypertrophy and angiogenic defects were observed.

miRNAs are required for lung development. Conditional deletion of Dicer in embryonic lung epithelium leads to inhibition of branching and increased apoptosis (Harris et al. 2006), suggesting that miRNAs play an important role in lung development.

miRNAs are required for lymphocyte development. (1) Deletion of Dicer at an early stage of T cell development compromised the survival of alphabeta lineage cells, whereas the numbers of gammadelta-expressing thymocytes were not affected (Cobb et al. 2005). In disease-free mice lacking Dicer in all T cells or harboring both Dicer-deficient and -sufficient T reg cells, Dicer-deficient T reg cells were suppressive, albeit to a lesser degree, whereas their homeostatic potential was diminished as compared with their Dicer-sufficient counterparts. However, in diseased mice, Dicer-deficient T reg cells completely lost suppressor capacity. Thus, miRNA preserve the T reg cell functional program under inflammatory conditions (Liston et al. 2008). (2) To explore the role of Dicer-dependent control mechanisms in B lymphocyte development, we ablated this enzyme in early B cell progenitors (Koralov et al. 2008).

miRNAs are required for cell-lineage decisions (Gu et al. 2008). (1) Using a transgenic line in which Cre recombinase is driven by the anti-Müllerian hormone receptor type-2 promoter, Nagaraja et al. (2008) conditionally inactivated Dicer in the mesenchyme of the developing Müllerian ducts and postnatally in ovarian granulosa cells and mesenchyme-derived cells of the oviducts and uterus. Deletion of Dicer in these cell types results in female sterility and multiple reproductive defects including decreased ovulation rates, compromised oocyte and embryo integrity, prominent bilateral paratubal (oviductal) cysts and shorter uterine horns. (2) A study analyzed function of miRNA biogenesis in germ cell development by using conditional Dicer-knockout mice in which Dicer gene was deleted specifically in the germ cells. Dicer-deleted PGCs and spermatogonia exhibited poor proliferation. Retrotransposon activity was unexpectedly suppressed in Dicer-deleted PGCs but not affected in the spermatogonia. In Dicer-deleted testis, spermatogenesis was retarded at an early stage with proliferation and/or early differentiation (Hayashi et al. 2008). (3) Mice lacking functional miRNAs in the developing podocyte were generated through podocyte-specific knockout of Dicer. These results suggest that miRNA function is dispensable for the initial development of glomeruli but is critical to maintain the glomerular filtration barrier (Ho et al. 2008; Harvey et al. 2008; Shi et al. 2008).

miRNAs are required for development and function of other organs and tissue (Suarez et al. 2008; Zhang and Nuss 2008; Zhao et al. 2007, Davis et al. 2008; Giraldez et al. 2005). (1) In mice, interference with miRNA biogenesis by tissue-specific deletion of Dicer revealed a requirement of miRNA function during limb outgrowth (Harfe et al. 2005) and in development of skin progenitors (Yi et al. 2006). (2) Ablation of Dicer in embryonic fibroblasts was found to upregulate p19(Arf) and p53 levels, inhibit cell proliferation and induce a premature senescence phenotype that was also observed *in vivo* after Dicer ablation in the developing limb and in adult skin (Mudhasani et al. 2008). (3) In newborn mice carrying an epidermal-specific Dicer deletion, hair follicles were stunted and hypoproliferative. These results reveal critical roles for Dicer in the skin and implicate miRNAs in key aspects of epidermal and hair-follicle development and function (Andl et al. 2006). (4) Retinal Dicer knock-out mice displayed a reproducible inability to respond to light (Damiani et al. 2008). (5) To test the requirement for Dicer in cell-lineage decisions in a mammalian organism, a conditional allele of dicer-1 (dcr-1) in the mouse was generated (Muljo et al. 2005).

13.5 Advantages and Limitations

After a number of applications to various conditions, Dicer inactivation technology has reached its mature stage for miRNA research, although it is bound by a few inherent limitations:

1. Dicer inactivation technology can be applied to study the global requirement of miRNA for certain biological and pathophysiological processes. But it does not allow for identification of key miRNAs to the processes; and does not even provide any information as to what particular miRNAs are involved in the processes.
2. The idea behind Dicer ablation is to interrupt biogenesis of miRNAs to reduce miRNA levels. That is, it is perceived by concept as a "loss-of-function" approach. However, in reality this aim may not be achievable in some circumstances. The opposite outcome, "gain-of-function", might be seen with this approach. For example, in a study reported by da Costa Martins et al. (2008), the authors failed to observe a preponderance of miRNA repression, despite efficient genetic Dicer deletion in the adult heart. While this may reflect the slow turnover or long half-life of certain endogenous miRNAs (Kim 2005), the observation that a subset of miRNAs was unexpectedly upregulated remained unexplained.
3. Dicer depletion can well result in lethality of animals rendering failure of further investigation. For instance, somatic Dicer loss results in embryonic lethality (Bernstein et al. 2003; Yang et al. 2005). Embryos or newborn mice lacking Dicer in developing heart succumb at embryonic day 12.5 or day 4 after birth with pericardial edema and a very poorly developed ventricular myocardium (Zhao et al. 2005; Chen et al. 2008).

The most important caveat regarding those experiments that disrupt miRNA biogenesis is the unknown extent to which other disrupted cellular functions contribute to the phenotype and to what extent the organism can circumvent a single gene knockout or knockdown.

References

Andl T, Murchison EP, Liu F, Zhang Y, Yunta-Gonzalez M, Tobias JW, Andl CD, Seykora JT, Hannon GJ, Millar SE (2006) The miRNA-processing enzyme dicer is essential for the morphogenesis and maintenance of hair follicles. Curr Biol 16:1041–1049.

Bernstein E, Kim SY, Carmell MA, Murchison EP, Alcorn H, Li MZ, Mills AA, Elledge SJ, Anderson KV, Hannon GJ (2003) Dicer is essential for mouse development. Nat Genet 35: 215–217.

Chen JF, Murchison EP, Tang R, Callis TE, Tatsuguchi M, Deng Z, Rojas M, Hammond SM, Schneider MD, Selzman CH, Meissner G, Patterson C, Hannon GJ, Wang DZ (2008) Targeted deletion of Dicer in the heart leads to dilated cardiomyopathy and heart failure. Proc Natl Acad Sci USA 105:2111–2116.

Chong MM, Rasmussen JP, Rudensky AY, Littman DR (2008) The RNAseIII enzyme Drosha is critical in T cells for preventing lethal inflammatory disease. J Exp Med 205:2005–2017.

Cobb BS, Nesterova TB, Thompson E, Hertweck A, O'Connor E, Godwin J, Wilson CB, Brockdorff N, Fisher AG, Smale ST, Merkenschlager M (2005) T cell lineage choice and differentiation in the absence of the RNase III enzyme Dicer. J Exp Med 201:1367–1373.

da Costa Martins PA, Bourajjaj M, Gladka M, Kortland M, van Oort RJ, Pinto YM, Molkentin JD, De Windt LJ (2008) Conditional dicer gene deletion in the postnatal myocardium provokes spontaneous cardiac remodeling. Circulation 118:1567–1576.

Damiani D, Alexander JJ, O'Rourke JR, McManus M, Jadhav AP, Cepko CL, Hauswirth WW, Harfe BD, Strettoi E (2008) Dicer inactivation leads to progressive functional and structural degeneration of the mouse retina. J Neurosci 28:4878–4887.

de Vries WN, Binns LT, Fancher KS, Dean J, Moore R, Kemler R, Knowles BB (2000) Expression of Cre recombinase in mouse oocytes: a means to study maternal effect genes. Genesis 26: 110–112.

Davis TH, Cuellar TL, Koch SM, Barker AJ, Harfe BD, McManus MT, Ullian EM (2008) Conditional loss of Dicer disrupts cellular and tissue morphogenesis in the cortex and hippocampus. J Neurosci 28:4322–4330.

Giraldez AJ, Cinalli RM, Glasner ME, Enright AJ, Thomson JM, Baskerville S, Hammond SM, Bartel DP, Schier AF (2005) MicroRNAs regulate brain morphogenesis in zebrafish. Science 308:833–838.

Gu P, Reid JG, Gao X, Shaw CA, Creighton C, Tran PL, Zhou X, Drabek RB, Steffen DL, Hoang DM, Weiss MK, Naghavi AO, El-daye J, Khan MF, Legge GB, Wheeler DA, Gibbs RA, Miller JN, Cooney AJ, Gunaratne PH (2008) Novel microRNA candidates and miRNA-mRNA pairs in embryonic stem (ES) cells. PLoS ONE 3:e2548.

Harfe BD, McManus MT, Mansfield JH, Hornstein E, Tabin CJ (2005) The RNaseIII enzyme Dicer is required for morphogenesis but not patterning of the vertebrate limb. Proc Natl Acad Sci USA 102:10898–10903.

Harris KS, Zhang Z, McManus MT, Harfe BD, Sun X (2006) Dicer function is essential for lung epithelium morphogenesis. Proc Natl Acad Sci USA 103:2208–2213.

Harvey SJ, Jarad G, Cunningham J, Goldberg S, Schermer B, Harfe BD, McManus MT, Benzing T, Miner JH (2008) Podocyte-Specific Deletion of Dicer Alters Cytoskeletal Dynamics and Causes Glomerular Disease. J Am Soc Nephrol 19:2150–2158.

Hayashi K, Chuva de Sousa Lopes SM, Kaneda M, Tang F, Hajkova P, Lao K, O'Carroll D, Das PP, Tarakhovsky A, Miska EA, Surani MA (2008) MicroRNA biogenesis is required for mouse primordial germ cell development and spermatogenesis. PLoS ONE 3:e1738.

Ho J, Ng KH, Rosen S, Dostal A, Gregory RI, Kreidberg JA (2008) Podocyte-specific loss of functional microRNAs leads to rapid glomerular and tubular injury. J Am Soc Nephrol 19:2069–2075.

Kanellopoulou C, Muljo SA, Kung AL, Ganesan S, Drapkin R, Jenuwein T, Livingston DM, Rajewsky K (2005) Dicer-deficient mouse embryonic stem cells are defective in differentiation and centromeric silencing. Genes Dev 19:489–501.

Kim VN (2005) MicroRNA biogenesis: Coordinated cropping and dicing. Nat Rev Mol Cell Biol 6:376–385.

Koralov SB, Muljo SA, Galler GR, Krek A, Chakraborty T, Kanellopoulou C, Jensen K, Cobb BS, Merkenschlager M, Rajewsky N, Rajewsky K (2008) Dicer ablation affects antibody diversity and cell survival in the B lymphocyte lineage. Cell 132:860–874.

Liston A, Lu LF, O'Carroll D, Tarakhovsky A, Rudensky AY (2008) Dicer-dependent microRNA pathway safeguards regulatory T cell function. J Exp Med 205:1993–2004.

Morita S, Horii T, Kimura M, Goto Y, Ochiya T, Hatada I (2007) One Argonaute family member, Eif2c2 (Ago2), is essential for development and appears not to be involved in DNA methylation. Genomics 89:687–696.

Mudhasani R, Zhu Z, Hutvagner G, Eischen CM, Lyle S, Hall LL, Lawrence JB, Imbalzano AN, Jones SN (2008) Loss of miRNA biogenesis induces p19Arf-p53 signaling and senescence in primary cells. J Cell Biol 181:1055–1063.

Muljo SA, Ansel KM, Kanellopoulou C, Livingston DM, Rao A, Rajewsky K (2005) Aberrant T cell differentiation in the absence of Dicer. J Exp Med 202:261–269.

Nagaraja AK, Andreu-Vieyra C, Franco HL, Ma L, Chen R, Han DY, Zhu H, Agno JE, Gunaratne PH, DeMayo FJ, Matzuk MM (2008) Deletion of Dicer in somatic cells of the female reproductive tract causes sterility. Mol Endocrinol 22:2336–2352.

O'Carroll D, Mecklenbrauker I, Das PP, Santana A, Koenig U, Enright AJ, Miska EA, Tarakhovsky A (2007) A Slicer-independent role for Argonaute 2 in hematopoiesis and the microRNA pathway. Genes Dev 21:1999–2004.

Rodriguez CI, Buchholz F, Galloway J, Sequerra R, Kasper J, Ayala R, Stewart AF, Dymecki SM (2000) High-efficiency deleter mice show that FLPe is an alternative to Cre-loxP. Nat Genet 25:139–140.

Shi S, Yu L, Chiu C, Sun Y, Chen J, Khitrov G, Merkenschlager M, Holzman LB, Zhang W, Mundel P, Bottinger EP (2008) Podocyte-selective deletion of Dicer induces proteinuria and glomerulosclerosis. J Am Soc Nephrol 19:2159–2169.

Sohal DS, Nghiem M, Crackower MA, Witt SA, Kimball TR, Tymitz KM, Penninger JM, Molkentin JD (2001) Temporally regulated and tissue-specific gene manipulations in the adult and embryonic heart using a tamoxifen-inducible Cre protein. Circ Res 89:20–25.

Srinivas S, Watanabe T, Lin CS, William CM, Tanabe Y, Jessell TM, Costantini F (2001) Cre reporter strains produced by targeted insertion of EYFP and ECFP into the ROSA26 locus. BMC Dev Biol 1:4.

Suárez Y, Fernández-Hernando C, Yu J, Gerber SA, Harrison KD, Pober JS, Iruela-Arispe ML, Merkenschlager M, Sessa WC (2008) Dicer-dependent endothelial microRNAs are necessary for postnatal angiogenesis. Proc Natl Acad Sci USA 105:14082–14087.

Wang Y, Medvid R, Melton C, Jaenisch R, Blelloch R (2007) DGCR8 is essential for microRNA biogenesis and silencing of embryonic stem cell self-renewal. Nat Genet 39:380–385.

Yang WJ, Yang DD, Na S, Sandusky GE, Zhang Q, Zhao G (2005) Dicer is required for embryonic angiogenesis during mouse development. J Biol Chem 280:9330–9335.

Yi R, O'Carroll D, Pasolli HA, Zhang Z, Dietrich FS, Tarakhovsky A, Fuchs E (2006) Morphogenesis in skin is governed by discrete sets of differentially expressed microRNAs. Nat Genet 38:356–362.

Zhang X, Nuss DL (2008) A host dicer is required for defective viral RNA production and recombinant virus vector RNA instability for a positive sense RNA virus. Proc Natl Acad Sci USA 105:16749–16754.

Zhao Y, Samal E, Srivastava D (2005) Serum response factor regulates a muscle specific microRNA that targets Hand2 during cardiogenesis. Nature 436:214–220.

Index

AMO. *See* Anti-miRNA antisense oligonucleotide or Anti-miRNA Antisense inhibitor oligoribonucleotides
AMO:miRNA interaction, 61, 170
AntagomiR, 36, 40, 128, 175
Anti-apoptotic miRNAs, 18
Anti-miRNA Antisense inhibitor oligoribonucleotides (AMO), 66
Anti-miRNA antisense oligonucleotide (AMO), 24, 68, 127–140
Apoptosis, 17–20, 22, 27, 28, 31, 87, 134, 137, 146, 172
Argonaute proteins (Ago), 4–6, 86
Arrhythmias, 25–27, 63, 138
Artificial intronic miRNA, 112, 115–123
ASO:mRNA interaction, 61, 164, 170

Brain cancer, 21
Breast cancer, 18, 21, 22, 38, 138, 146, 149

Cancer, 10, 11, 17–23, 27, 32, 37, 38, 63, 87, 99, 108, 123, 137–139, 146–149, 158
Cardiac hypertrophy, 17, 23–25, 138
Cardiovascular disease, 23–28
Colon cancer, 18, 22, 23, 87, 123
Complementarity, 6–8, 84, 93–95, 99, 157, 168
Conventional antisense technique, 164
Cre-loxP, 120, 121, 177, 185

Derepression, 60, 61, 136, 170, 180
Development, 2, 8, 12, 16, 17, 19–21, 28–29, 31, 32, 35, 37, 38, 40, 71, 99, 108, 112, 122–124, 137, 158, 175, 183, 186, 187
DGCR, 8, 4, 5, 70, 183
Dicer, 5, 12, 16, 27, 28, 32, 34, 40, 60, 67, 70, 71, 75, 76, 78, 118, 128, 183–188
Dicer inactivation, 71, 183–188
DNA methylation, 36, 37
Drosha, 4, 28, 34, 40, 70, 128, 136, 157, 183
Drug design, 67, 69–71

Epigenetics, 36–38
Expression, 6, 8–12, 16–24, 26–28, 30–41, 59, 60, 64, 66, 69, 70, 76–81, 85, 87, 97, 99, 105, 107, 111–113, 116, 120–124, 128, 133, 135, 136, 138–140, 146, 149, 154, 156–158, 162–166, 169, 170–172, 176, 179–181, 183–185

FLP-FRT deletion, 177–179
Fragile X syndrome, 29, 121

Gain-of-function, 12, 16, 41, 60, 66, 67, 69, 75, 87, 101, 111, 122, 138, 188
Gene silencing, 6, 26, 41, 59, 98, 102, 108, 112, 120, 122, 124
Gene-specific, 61, 66, 67, 94, 95, 97–100, 105–107, 138, 162–166, 171
GFP, 103, 104, 107, 108, 154–156, 158
Glucose homeostasis, 35, 36

Hairpin, 4, 5, 66, 69, 75–78, 83, 101–109, 112, 119, 122, 124, 130, 131, 137
Heart failure, 9, 17, 23–25
Histone modifications, 36, 37
Homologous recombination, 79, 176–180, 184, 185
Host miRNAs, 32–34

In situ hybridization (ISH), 81, 84–86
Intergenic miRNAs, 2, 3, 122
Intragenic miRNAs, 2, 4
Intronic miRNAs, 2–4, 112, 115–123, 176

Knockdown, 18, 22–24, 28, 36, 41, 59, 60, 70, 77, 78, 94, 102, 103, 107, 108, 124, 128, 135, 138–140, 146, 147, 168, 175, 182–184, 188
Knockout, 16, 31, 34, 39, 41, 60, 66, 67, 70, 71, 109, 124, 139, 146, 153, 175–185, 187, 188

Lentivirus vector, 102–104
Leukemia, 9, 18, 20–21, 123
Lifespan, 34, 35
Lipid metabolism, 36
Locked nucleic acid (LNA), 77, 131–133, 148, 169
Loss-of-function, 12, 27, 35, 41, 60, 61, 67, 70, 128, 136, 138, 153, 154, 158, 162, 170, 175, 180, 186, 188
Luciferase, 78–80, 96, 103, 105, 107, 135, 149, 156
Lung cancer, 10, 18–21, 123
Lymphoma, 20, 22–23, 31, 32, 123

Mature miRNA, 4, 5, 10, 22, 28, 66, 75, 77, 80, 83, 85, 112, 113, 118, 120, 121, 128, 129, 131, 136, 139
Metabolic disorders, 34–36, 138
Microinjection, 114
MicroRNA (miRNA), 1–41, 59–71
miRISC, 5–7, 86, 93, 120, 135, 157
miR-KO. *See* miRNA Knockout
miR-Mask. *See* miRNA-masking antisense oligonucleotides
miR-Mimics. *See* miRNA Mimics
miRNA. *See* MicroRNA
miRNA biogenesis, 12, 32, 40, 70, 78, 183–187
miRNAi, 39, 59
miRNA interference technologies, 59–71
miRNA Knockout (miR-KO), 66, 71, 139, 175–183
miRNA-masking antisense oligonucleotides (miR-Mask), 66, 67, 70, 71, 158, 161–173
miRNA Mimics, 66, 69, 70, 76, 77, 80, 93–101, 105–108
miRNA:mRNA interaction, 6, 7, 35, 60, 61, 64, 163, 164, 171
miRNA seed family, 64–65, 67, 153, 154, 156–158, 168, 169, 171
miRNA-specific, 60, 61, 67, 70, 83, 128, 136, 140, 158, 161, 163, 164, 170
miRNA sponge, 66, 67, 140, 146, 153–158, 161, 168–172
miRNA-targeting, 60, 61, 66, 67, 101, 102, 111, 112
miRNA transgene, 66, 70, 111–124
MT-AMO. *See* Multiple-target anti-miRNA antisense oligonucleotides
Multi-miR-Mimic. *See* Multi-miRNA Mimics
Multi-miRNA Hairpins, 66, 69, 101–109
Multi-miRNA Mimics, 66, 70, 101–109
Multiple-target anti-miRNA antisense oligonucleotides (MT-AMO), 66, 140, 145–150, 154, 157, 158, 161, 162, 168
Mutation, 2, 8–12, 29, 35, 84

Neomycin (Neo), 176, 177, 179, 184–185
Neuronal disease, 28–29
Northern blot, 75, 80–83, 115, 135

One-drug, multiple-target, 63–64, 148, 149

PCR, 76, 83, 103, 104, 113–115, 156, 176, 177, 179, 184, 185
Pol II, 4, 103–108, 122
Pol III, 3, 77, 78, 120, 156
Polymorphism, 10–12, 166
Precursor miRNAs, 4, 76, 102, 177, 183
Primary miRNAs, 3, 78
Proapoptotic miRNAs, 18
Probe, 21, 82–86, 134, 184

qRT-PCR. *See* Quantitative reverse transcriptase-polymerase chain reaction
Quantitative reverse transcriptase-polymerase chain reaction (qRT-PCR), 80, 83, 156, 157

RISC. *See* RNA-induced silencing complex
RNAi. *See* RNA interference
RNA-induced silencing complex (RISC), 5, 6, 29, 33, 59, 76, 128, 131–133, 135, 136
RNA interference (RNAi), 28, 29, 32, 38, 39, 59, 60, 94, 121, 123, 128, 135, 154
RNA oligonucleotides, 84–86

SC-miRNA. *See* Synthetic canonical miRNA
Seed Site, 6, 7, 10, 64, 65, 77, 93, 94, 96, 99, 140, 147, 148, 167–170
Short hairpin RNA (shRNA), 38, 76–78, 113
siRNA. *See* Small interference ribonucleic acids
Small interference ribonucleic acids (siRNA), 6, 28, 38, 40, 60, 69, 76–78, 88, 94, 95, 97–100, 108, 138, 139, 150, 184
Sponge miR-Mask, 66, 67, 71, 167–173
SpRNAi, 115–119
Synthetic canonical miRNA, 76–80, 86–88, 98, 99, 101, 108, 156

Target gene, 8, 10, 11, 17, 21, 22, 25, 26, 35, 38, 59, 61–65, 67, 70, 71, 80, 86, 95, 97, 100, 103, 105–107, 120, 122, 123, 132, 135, 136, 138, 140, 147, 149, 157, 158, 162–164, 167–172, 181
Targeting-miRNAs, 60, 61
Target protector, 162, 164, 165
Tourette's syndrome, 11, 29
Transgenic mice, 24, 66, 75, 107, 114, 115, 123, 177, 179, 185

Vascular angiogenesis, 27–28
Viral disease, 30–34
Viral miRNAs, 30–33